THE WATCHMAKERS of MASSACHUSETTS

August C. Bolino

Kensington Historical Press
Box 1314 Cardinal Station
Washington, D.C. 20064

ISBN 0-939133-01-6

Printed in the United States of America
by
BookCrafters
Chelsea, Michigan

Books by August C. Bolino

The Watchmakers of Massachusetts (1987)

The Ellis Island Source Book (1985)

Career Education: Contributions to
Economic Growth (1973)

Manpower and the City (1969)

The Development of the American Economy (1966)

The Development of the American Economy (1961)

TABLE OF CONTENTS

LIST OF TABLES

LIST OF FIGURES

PREFACE

If you ask me why Massachusetts, there are two simple answers. The simplest is that I was born there. The other I need to explain, because I came to watch collecting by a roundabout fashion. I graduated from Mechanic Arts High School (now Boston Technical), and I worked as a machinist at the Chelsea Clock Company. This work was interrupted when I left the state in 1942 to enter the Army Air Force. From then until 1970, my hobby was dormant. Then I saw an advertisement in a British antique magazine that offered two pounds of watches (weight) for a British pound (money). I could not resist this bargain. After the box arrived, I tinkered with this junk for a while, and when I got one of the watches to run, I was hooked.

At first, I rambled all over the horological map. I made the mistake of beginners: I collected everything. But eventually, I returned to my roots: I began to specialize in Massachusetts. But it was a mammoth task. What to collect? After all, the state produced over 40 million watches. I chose to limit myself to types by company. This continuing quest led me to a surprising fact. Noone had ever written a book on the watchmakers of Massachusetts. As an economist, I had to go about earning my daily bread, but I started collecting information by company, knowing that some day I would put it all together.

I have written books before, but this time there was an incredible amount of technical information to absorb. Having gone through it, I have much more respect for those who made my path easier (Crossman, Abbott, Moore, Ehrhardt, Townsend, Harrold and others).

My main research debts are to the Library of Congress, the Lending Library of the National Association of Watch and Clock Collectors, Columbia , Pennsylvania and to Sarah Hamrick, of the

Catholic University of America, where I teach. She labored much to find a disparate array of sources when she "manned" the interlibrary loan desk. I should like ,also, to thank Nick Crettier, who photographed my own watches; an anonymous referee, who made excellent comments about Chapter 2 and my wife, "TJ," who helped with the layout of the manuscript.

August C. Bolino
Washington, D.C.
January 1987

Tho. Sargeant
Clock & Watch
MAKER,
SPRINGFIELD.
Huntington & Church
Watch-Makers
Springfield,
Massachusetts.
Caleb Rogers.
Clock & Watch
MAKER,
At the Dial
NEWTON.
A Willard
Watch & Clock
MAKER
BOSTON
A.C.TROTT & Co.
Watch & Clock Makers
No 28 Marlboro St
BOSTON.

INTRODUCTION:
A BRIEF HISTORY OF THE MASSACHUSETTS WATCH INDUSTRY , 1853-1957

The development of the American System of Manufactures meant that in the United States watches could be produced in a single plant by the thousands, whereas in Europe watches were still produced in parts in different locations and assembled elsewhere. The American approach resulted in part from the relative scarcity of labor. Standardized parts economized on labor and repairs. In addition, the industrial progress in the United States was tied to a system that moved labor to the raw materials, rather than vice versa, as was customary in the older manufacturing centers of Europe. (1) Thus, while American labor was more expensive, when combined with equipment that was more productive, it enabled our producers to provide an inexpensive timepiece for the population.

The industrial revolution helped to make petty capitalists (artisans, farmers and storekeepers) into industrial capitalists. These industrialists perfected new productive techniques and learned how to market their increased output. The history of the Massachusetts watch industry illustrates this transformation. These watch companies adopted policies which were needed to survive, but they suffered from the normal development of industries which usually go through stages of rapid growth, increased competition and decline.

In Europe, in 1850, watch manufacturing was controlled by craft guilds, but in the United States, where guilds never flourished, the factory system was dominant. In short time, the American system of manufactures provided the growing population with guns, sewing machines, farm implements, clothes and clocks and watches.

Harrold classifies the growth of the United States watch industry in three stages:

```
    1. conventional (expensive)watches;
    2. inexpensive jeweled watches and
```

3. dollar watches. (2)

As is their wont, the Americans demanded a very inexpensive watch, which demand could only be provided by mass production. Thus it was that between 1890 and 1930, millions of dollar watches were produced. This meant, also, that in Massachusetts, as elsewhere, those companies making conventional watches had a very difficult time staying solvent. Many companies came and went as competition became fierce between 1880 and 1930.

The beginning of American superiority was the survival of the American Watch Company at Waltham, Massachusetts. With Royal E. Robbins as President and Aaron Dennison as Superintendent, the company succeeded where its predecessors had failed. The Civil War was a stimulant to success, with the increased demand for military watches. Brearley states that this put "the Waltham Company squarely upon its feet by justifying quantity production." (3) The first dividend was declared and paid in 1860, and by 1864, the company had produced 118,000 watches. By that year, the Ellery, the military watch, accounted for nearly one-third of total sales. (4) More important to the company, the war provided capital for growth and expansion.

As Schumpeter said, success breeds imitators. (5) When the Boston Watch Company failed in the depression of 1857, Edward Howard returned to Roxbury to organize the E. Howard & Company, whose name was changed to the Howard Clock & Watch Company in 1861. In that same decade, we saw the formation of the National Watch Company (Elgin) and the Illinois Springfield Watch Company in 1869. To quote Landes,"These and others were to produce over a period of three quarters of a century about one hundred and twenty million jeweled watches plus an unknown number of so-called dollar watches or clock- watches." (6)

The Elgin Company was created by a group of western

capitalists, at the suggestion of a few men who had been trained at Waltham. This company was an immediate success, and it was able to pay a dividend from the beginning. It produced the astonishing total of 50 million watches--almost 10 million more than Waltham.

The success of the American watch industry was linked also to the size of the market. In 1850, the average American could not afford to purchase a watch, but in the second half of the century, more goods were made for home consumption. Our population expanded from natural causes and from immigration, and this provided the largest consuming unit (market) in the world. As such, our producers had little incentive to seek a wide foreign market. Therefore, our manufacturing efforts were not dependent on the vagaries of world economic conditions, particularly during foreign political upheavals and wars.

While tariffs increased during most of the nineteenth century in the United States, the watch industry did not need protection. During the Civil War, several tariff laws were enacted because of financial necessity, but when fighting ceased, legislation ceased also. But the protective tariff was a central issue of the time. As tariff rates were raised, a peculiar problem arose: the revenue from customs duties exceeded the meager financial demands of the federal government. As President Grover Cleveland said later, "It is a fact that confronts us not a theory." One consequence was the lowering of duties in 1872 by 10 percent, but they were restored by the outgoing Republican Congress in 1875.

But 1872 was important for another reason. In that year, George F. Ballou and John E. Whitcomb began to manufacture watch repairer lathes, later known as the "Whitcomb Lathes." When Ballou retired in 1874, Whitcomb joined with Henry Fisher and Ambrose Dexter to form the American Watch Tool Company, which made lathes and watch machinery. Webster died in 1893 and

Whitcomb in 1906, and in 1918 the company was liquidated. It became a part of the U. S. Watch Company of Waltham, which in turn became part of the Keystone and Howard Companies.

The nation's centennial was the first international exhibition ever held on this continent. American producers were supposed to learn from their European conterparts, but in fact, they were astounded at America's progress in producing interchangeable parts. Between 1870 and 1875, the number of imported Swiss watches fell from 330,000 to 134,000, and American suppliers were sending their products to Europe. (7) One foreign observer commented," I sincerely confess that I personally have doubted this competition. But now I have seen it-- and I have felt it--and I am terrified by the danger to which our industry is exposed." (8) His fears were justified: in 1878, the average output per worker in Switzerland was 40 watches, compared to 150 watches at Waltham. (9) This increased productivity helped to reduce the price of an average watch in the United States from $40 in 1850 to $10 in 1880, and the inexpensive non-jewelled watches could be obtained for one dollar in 1900.

Some of these results were surprising, because the American economy was in the midst of a six-year depression. It began with overinvestment in railroads and finished with the collapse of the banking system. When Jay Cooke and Company suspended payments, other financial houses followed. The collapse spread soon to textiles and farm products at home and to other industries abroad. Trade and industry continued stagnant until 1878, and the number of business failures doubled. (10) This was an era of falling prices and diminished profits. The economic climate was worsened by continuous labor strife (especially after the railway strikes of 1877).

In this time, Massachusetts added three more watch companies. In 1875, the Fitchburg Watch Company was organized, but it was discontinued for lack of funds. In 1877, the Hampden Watch

Company was organized in Springfield, Massachusetts; it took over the business of the New York Watch Company. Two years later, the Auburndale Watch Company was started; it purchased the machinery of the U. S. Watch Company Company at Marion, New Jersey. The formation of the Auburndale Company goes back to the work of Jason R. Hopkins, of Washington, D. C., who in 1875 perfected a watch that he intended to sell for 50 cents. He linked up with Edward A. Locke, of Boston, and W. D. Colt, also of Washington, D. C. It was Hopkins who got the financial support of W. B. Fowle, of Auburndale, Massachusetts. Fowle put up $250,000 in this venture, but the watch was a failure. (11)

In the second half of 1879, prosperity was renewed. Our exports exceeded imports, textiles recovered, railroad earnings rose, and agricultural output more than doubled. By the end of the decade, the seven-year panic was over. These improvements might have influenced Benedict and Burnham in founding the Waterbury Company in 1878. In that year, D. A. Buck obtained his patent for an inexpensive, machine- made watch with only 58 parts. It had a long mainspring that rotated like a touribillon, carrying the minute hand with it. The dial was of paper and the escapement duplex. This watch was discontinued in 1891 and was really replaced in the next year by the Ingersoll model. But the Idea of a "cheap" watch originated in Switzerland, where in 1867 Roskopf produced a pin pallet lever watch. It was exhibited at the 1867 Paris Exposition. He used few parts, a three-wheel train, keyless winding and a finger adjustment for the hands. (12) This was the beginning of the large numbers of pin-pallet watches that were sent to the United States in the first half of the twentieth century, and it posed a new threat to Massachusetts watch producers.

The year saw another of the periodic downturns in economic activity. This was considered normal after four or five years of recovery, but the contraction was widespread. Pittsburgh steel felt the decline, as did textiles and railroads. Between 1879 and 1883, the United States put down over 40,000 miles of railroad tracks. In the panic of 1883, the price of these rails fell from $70 per ton in 1879 to $33 in 1883. (13)

In 1879, the United States was a long way from being the leading producer of clocks and watches. For total output, France led the list in that year. For watches alone, the Swiss produced 1,500,000 pieces, the French 500,000, the United States between 300,000 and 350,000 and the British 200,000. (14)

In the economic recovery, that commenced in 1886, the English decided to strike back at American watch producers by attempting to adopt their methods. The Lancashire Watch Company, of Prescott, commenced in 1888 to buy up all the tools and watch parts to minimize outside competition and to obtain resources to mass produce a watch. The effort failed, because old watchmakers rebelled at the new techniques that were forced upon them. Therefore, the British did not succeed in making complete watches in a single factory, as Waltham, Elgin and other American companies did.

Charles S. Crossman, writing in 1888, noted that American watches were superior. In his words, "There are practical reasons why the American watch, and especially the Elgin, is the best and and why it sells the best. The American will, as a rule, stand more hard usage and still keep good time; its exposed parts may not be so elaborately finished as some grades of Swiss watches of comparatively the same commercial value, but this is more than counterbalanced by the temper of the steel parts, the close adjustment to temperature, the interchangeability of the parts of the movement, and the ease with which they are procured for repairing purposes." (15)

We mentioned above that the United States constructed more

than 40,000 miles of railroad track by 1883. The growth of the watch and the railroad industries ran as parallel as the tracks, because the trains required good timepieces. This became evident when two trains of the Lake Shore Railroad collided in 1891. One of the watches had acquired dust and soot and was four minutes off, hence the collision. This aroused interest in a "railroad" watch that would meet more exact standards than existed at the time. In fact, the Hamilton Watch Company was organized in 1892 to meet these new standards.

Once again, in 1893, a depression intruded on the doings of the Massachusetts watch industry. This downturn was severe, causing an abrupt cessation of demand for consumers goods (including watches). The financial strain was protracted and prices fell by 13 percent. Between 1893 and 1896, gold left the country, foreign investment in the United States dried up and securities prices plummeted. The market for watches fell and Waltham Watch Company shut down.

The presidential election of 1896 settled several questions concerning gold and silver, free trade and protection and the general direction of the economy. A new confidence turned into optimism and the United States economy started a long upward move from a period in which it suffered three serious downturns in 20 years (1873, 1884, 1893).

In 1899, the watch industry was jolted by an antitrust case against the Keystone Watch Company. The Sherman Antitrust Act, passed in 1890, declared that combinations in restraint of trade are illegal, and that anyone who attempts to create a monopoly, is guilty of a misdemeanor. Although this suit involved the "filled watch case industry only," it interests us here because Keystone, at the time, controlled the E. Howard Watch Company and the United States Watch Company. (16) The company that was sued began as a merger of the T. Zurbrugg Company (which was acquired by the Riverside

Watch Case Company, of Riverside, New Jersey) and the Keystone Watch Company, of Philadelphia. This parent company created a subsidiary, the Philadelphia Watch Case Company, to give the impression that the two companies were operating independently. The subsidiary "reached backward" to acquire the New York Standard Watch Company in 1901.

In 1903, the Keystone Company obtained in bankruptcy the trade marks and good will of the E. Howard Clock Company and the common stock of the Crescent Watch Case Company, which had purchased in 1890 the entire case business of the American Waltham Watch Company. The tangled financial web continued in 1906, when Keystone acquired 42 percent of the American Watch Case Company of Toronto (the remaining 58 percent was owned by Waltham and Elgin). Keystone created the Keystone-Crescent Watch Case Company, of Canada, and it proceeded to sign a contract with the Elgin Watch Company in 1904, which gave Elgin control of all business except in Canada. This was followed in 1909 with a contract with Waltham Watch Company, in which Keystone was made the sales agent for all cases in all contries, except Great Britain, France and Spain.

In a circular sent to all jobbers by Keystone, the company spelled out the terms under which Howard watches were to be sold. The letter stated that no movement shall be removed from a case and no watch shall be sold to any supplier who was "objectionable to the manufacturer." The company also specified a minimum price for the watch, below which retailers "shall not advertise, sell or offer." (17)

Between the turn of the century and the end of World War I, there were not many technical improvements in producing watches. There were, however, developments in metallurgy. The problem of changing temperatures plagued the watch makers for years. The answer was to produce a balance wheel of a combination of materials that would "compensate" for temperature changes. In 1900,

Guillaume used a nickel-iron alloy instead of steel. When this alloy was combined with brass, it eliminated the middle-temperature error (the error that remained after the inward and outward motions of the balance wheel were compensated). By 1919, Guillaume produced an alloy--Elinvar--that did not change with temperature.

As usual, the prosperity that started in 1896 was interrupted by a recession. Several industries affiliated with steel established new highs, but the downturn in the spring was once again precipitated by a selloff in railroad securities and a fall in the price of steel. But in this recession, money was as scarce as ever, leading historians to call it "The Money Panic of 1907." This panic, more than any other factor, led to the creation of the National Monetary Commission and the passage of the Federal Reserve Act of 1913.

The period after World War I was one of intensified foreign competition. The United States, Switzerland and Germany produced the largest share of watches and clocks. By 1925, the value of watches and clocks manufactured in the United States almost equalled the combined production of the other two. American output in 1925 was two and one-half times what it was in 1914. (18) In this time, exports of watches and parts grew by only $218,000 ($1,460,424 to $1,678,952), while imports rose by $10.7 million (from $3,386,738 to $14,162,840). There where interesting shifts in the pattern of trade. In particular, the market for American watches shifted from Europe to the Far East. In 1913, the value of exports of watches and parts to Europe totalled $705,473, compared to $49,713 to the Far East. By 1926, the Far East total was $555,145 and that of Europe only $208,756. In terms of single countries, however, Canada and Australia ranked 1-2 in both 1926 and 1927. In fact, exports to Australia increased 12 times between 1913 and 1927. (19)

The above export figures were an insignificant part of total United States production, but for Switzerland the watch industry was dependent on exports. The production of watches began in Switzerland in the Canton of Geneva in the sixteenth century, but the real development of Swiss watches began with the arrival of French refugees after 1685. With the revocation of the Edict of Nantes, a number of famous French watchmakers settled in Geneva. By contrast, the watchmakers at Neuchatel were locally trained, mainly at Le Locle. These two areas supplied most of the 62,833 watchmakers in Switzerland in 1920 and most of the export of watches.

During World War I, there was overproduction in the United States and in Switzerland, leading to price cutting and "chablonnage" (sending unfinished parts overseas to be made into movements). Although the United States had an advantage of close ties to the railroad industry and a better system of buying agents, the Swiss still managed to account for one-half of all watches sold in the United States in 1920.

By the year 1900, the binge in the formation of new companies as over. As the public demand for jeweled watches declined, it was more difficult for a watch company to remain a viable financial entity. Companies began merging or disappearing in bankruptcy. By 1900, there was a new challenge from the Swiss, who had learned how to cope in United States markets. The Swiss learned American technology and marketing techniques. More important, they began to invade the United States with their small attractive, efficient wrist watches, first for the ladies, later for men.

In accordance with trends of the time, and to meet growing foreign competition, the Swiss attempted in 1927 to cartelize the industry. Ebauches were set at a fixed price, production was controlled by Ebauches, S. A., and tariffs were established. In 1928, the Swiss Watchmaking Credit was created to operate as an intermediary between banks and watch-making companies. These economic actions posed a very serious threat to the continued prosperity of the Massachusetts companies. (20)

Besides Swiss competition, the Massachusetts watch industry was damaged by the development and mass acceptance of the wrist watch. Like the pocket watch, the wrist watch was a European development. Before wrist watches could be mass produced, a technique was needed to produce miniature movements. Gohl tells us that David Rousseau, of Geneva, made a verge fusee only 18 mm in diameter "in the late 1600's." (21) In time, others made smaller movements. By 1896, a 6.75 mm watch movement was exhibited at the 1896 Geneva Exhibition.

The first wrist watches grew out of watches "to be fixed to a bracelet." But here, once again, military necessity propelled the watch industry. In the 1880's, a German officer discovered that he was unable to hold a pocket watch and operate efficiently in the field. He recommended attaching the watch to the wrist (which meant reducing its size). The Germans ordered these watches and the Swiss reacted. But the first wrist watches shipped to the United States at the beginning of the twentieth century could not be sold.

When the Americans began to experiment with wrist watch production after 1907, the women wore them as jewelry, but the men rejected them. In 1911, the Dueber Watch Case Company manufactured wrist watch cases that were later joined with the Hampden movements. Waltham claimed, in that year, that the 10 ligne movement (about the size of a nickel) was the smallest watch in the world that could be mass produced. In the following year, Gruen advertised its new "Wristlet" watch in 26 different models.

In World War I, both the United States and Switzerland produced wrist watches in quantity, some with bold numerals and luminous hands for viewing in the trenches at night. In 1919, Waltham declared that the wrist watch was "no longer a fad or an eccentricity--it is an essential of every-day life." (22)

While all the original wrist watches had round movements, the prosperity decade of the 1920's brought a new demand for fashion. Producers offered rectangular, square, oval, baguette and tonneau shapes. The unusual designs were encouraged by the Art Deco Exhibit of 1925 in Paris.

By 1927, wrist watches outnumbered pocket watches in the United States for the first time. The loss of popularity of the waistcoat (vest) contributed to the decline. By 1930, the ratio of wrist watches to pocket watches produced was 50 to 1. (23) By then, the winding stem was at 3 o clock, bezels were snap type or hinged, watch "glasses" were of plastic (which yellowed with age), and the cylinder had given way to the lever movement.

But these wrist watch improvements came at a bad time. As the stock market crash reverberated throughout the economy, several watch companies failed or were absorbed. The New York Watch Company went down in 1929 and Hampden in 1930. The Illinois Company was taken over by Hamilton in that year and Howard the next year. In 1933, the South Bend Company collapsed.

The American watch industry was clearly in trouble. Consumer purchasing power was very low, as the Gross National Product fell by 50 percent, and the American suppliers could not meet the competition of new Swiss styling in wrist wtches. Thus, in the depression, the American watch industry suffered from serious overcapacity. The Congress tried to help by passing the Smoot-Hawley tariff, but it is significant that over six thousand economists sent a telegram to President Hoover urging him not to sign the bill. They knew that foreign trade is a two-way street and that the restriction of imports would affect exports. The first effect of the high tariff on Swiss watches was a large rise in smuggling. Frederic Dumaine, the

President of the Waltham Watch Company, estimated that one to two million movements were smuggled into the United States, which cost the United States $2-5 million in lost revenues. (24) In addition, when smugglers were apprehended (which was not often), the watch movements that were seized were sold at auction-- thereby disadvantaging American producers a second time. The Swiss blamed the tariff and pointed out that it exceeded the value of the movements being smuggled. But in spite of their complaints the Swiss share of total United States sales grew steadily.

American producers petitioned President Roosevelt many times for redress of this economic dilemma, but the President was bent on nursing his free-trade policies--that he ordered Secretary of State Hull to carry out. The initial result was a 34 percent cut in tariffs on jeweled watch movements. (25) The Americans were forced to find private solutions to their competitive problems.

In a classic example of the "if you cant lick em, join em" case, the Americans accelerated a practice that they had restarted after World War I, that is to import parts and to assemble watches in the United States. This practice had declined when the tariff was raised, but after the New Deal cuts, the Americans found themselves competing with lower foreign wages, more aggressive sales techniques and a government-protected industry in Switzerland. The Swiss had created a cartel to avoid destructive competition and to encourage stability in price and output.

The American answer was the National Industrial Recovery Act, passed on June 16, 1933. It's primary purpose was "to eliminate unfair competition and overproduction, to restore normal profits to business, and to provide a living wage to laborers." (26) The law established Codes of Fair Competition, and the watch code was approved as early as August 1933. It raised wages in the watch industry by about 25 percent. It should be obvious that this was not a solution to what troubled the American watch industry.

The NIRA was declared unconstitutional by the United States Supreme Court in *Schechter Poultry Company v United States* as an improper delegation of legislative authority. During the case, the Congress was considering a permanent law to guarantee workers the right to organize and to bargain collectively. When the NIRA was set aside, the National Labor Relations Act (the Wagner Act) was passed on July 5, 1935. The key clause was old Section 7a of the NIRA, which became Section 7 of the Wagner Act. It forebade employers from interfering with employees in the exercise of their rights to join unions, to select representatives of their own choosing and to bargain collectively.

The effect of the Wagner Act on the Massachusetts watch industry can be studied in the case of Waltham. In 1924, after 75 years of industrial labor peace, Waltham was shut down by a strike. It stemmed primarily from a wage cut in troubled financial times. This history must have weighed heavily on management. Dumaine accepted the NIRA readily, although his management philosophy was diametrically opposed. The workers were granted a 40-hour week, and in 1935, profits reappeared. In this climate, the Company did not oppose the organization drive in 1940 of the International Watch and Jewelry Workers Union. The union won representation by a vote of 1,510 to 496.

World War II saved the American watch industry temporarily. It produced aircraft clocks, chronometers, compasses, time fuses for bombs, speedometers and navigational time pieces. These items were produced in quantity, and they achieved an unusual standard of excellence. Bu 1943, for example, Waltham watches gained or lost an average of only 12 seconds per day--better than the required government standards. (27)

At the end of World War II, the Massachusetts watch industry appeared to be in sound financial health; Waltham had $10 million of unfilled orders at war's end. But the production victors in the post-war period were those who adjusted quickly to a number of technological revolutions in watchmaking. The first major change was the availability of quality wrist watches at reasonable prices. These watches had automatic winding. The chronographs added date, alarm and waterproof and shock resistant movements. These developments were mainly Swiss; in the United States, technological improvements concentrated on electrical applications. In 1952, the battery-powered watch became available. In the electronic watch, the balance spring and wheel are replaced by electronically-induced vibrations of a tuning fork or by a crystal vibrator.

But it was really Timex Corporation that revolutionized the marketing of watches. These simple and inexpensive timepieces were disposable. The plates were attached automatically, hence the watches could not be disassembled. When one stopped functioning the "repairer" simply replaced the entire movement, or the consumer disposed of the watch and purchased another.

These developments are best mirrored by the demise of the Waltham Watch Company. Waltham, like most watch companies, had been in trouble in the 1930's, but World War II returned it to profitability and the Korean War offered another temporary stimulus. Between 1950 and 1953, expenditures for national security rose from $18 billion to $52 billion, with aircraft and electronics important elements. (28) But Waltham's troubles reappeared soon, and the company was sold. We can ask, " Who (or What) Killed Waltham?" (29) The beginning of the end occurred on December 28, 1948, when Waltham filed an application for reorganization in the District Court of Boston. The issue became a national one: should the Federal government allow the failure of a company which had contributed mightily to the war effort and which employed one-fifth of all the workers in Waltham, Massachusetts? A second factor was the administration of Frederic C. Dumaine. He became chief executive in 1922 when Kidder, Peabody and Company refinanced the company. He invested $125,000 of his own money, and he began a ruthless program to bring the company back to profitability. He reduced wages and installed a new system of factory discipline. By the end of World War II, he had achieved his objective. Waltham was a highly profitable business, and at this point Dumaine cashed in. He paid off the long term bonds, he reduced the preferred stock by 78 percent, and he sold his equity at a substantial profit. Paying off the bonds compounded the reconversion problems, because it sacrificed the needs for new plant and equipment.

A third factor was the decline of quality. Consumer and jewelers complaints multiplied. The 670 and 675 movements were often dirty and they were in cases that did not fit. As conditions worsened, President Ira Guilden quietly sold all of his shares. A second executive left the sinking ship.

The new chief, Paul P. Johnson, commenced a large advertising campaign to celebrate the "33,000,000th watch." He spent more on advertising in one year than Waltham had in the previous 25 years. The advertising campaign also featured a Centennial series, in anticipation of the 100th birthday of the company. The company promised a watch with a mainspring that was "guaranteed to last until 2050 AD."

Waltham made several bailout attempts. In April 1949, it was granted a $6 million loan from the Reconstruction Finance Corporation--one third to be used for plant modernization. The company sold its large inventory at one-half price to improve its cash position. But by February 3, 1950, the company was shut down again. After a new reorganization was approved by the courts,

Waltham moved on another front: the limitation of imports. The number of imports grew substantially under the 1936 agreement of the Tariff Commission. Waltham, Elgin and Hamilton urged President Truman to curtail imports from Switzerland. When the Swiss pointed out that the Americans could not possibly supply the entire market and that Benrus, Bulova and Gruen employed nearly 4,000 skilled technicians in this country, Truman turned down the request for a 50 percent increase in tariffs on Swiss watches.

Truman's action called for a new Waltham strategy: it would import 17-jewel movements, and it would concentrate on producing 19- and 21-jewel movements. The trustees of the Waltham Watch Company announced that this decision was thrust upon them by the great disparity in wages (American wages were two and one-half times greater than the Swiss).

In 1953, a new factor was introduced: the allegation of a "Swiss Watch Cartel." The Department of Justice claimed (belatedly) that the Federation of Watch Manufacturers in Switzerland violated the Sherman Antitrust Act. This new interest in antitrust action may have been prompted by the attempt of Benrus to obtain the Hamilton Watch Company.

None of the above actions offered much hope of saving Waltham. The company had six different managements between 1947 and 1957. When Bellanca finally purchased Waltham in 1957, the watch company ceased to be a viable entity. The company name was changed to the Waltham Precision Instrument Company. For a while, it continued to sell watches and clocks, but only to retain its tax-loss credits. This ignominious end closed a century of watch production in Massachusetts.

As the Massachusetts watchmaking industry faded away, we could lament this passing, but as Harrold reminds us, Americans continued to make important innovations in this industry until quite recently. The electric watch was introduced in the 1950's and the electronic watch in the 1960's. These have been replaced by the throwaway solid state watches that can now be purchased for as little as $2.50 (30) The pendulum has surely swung back to that most American of characteristics: we always prefer low-priced goods. But as collectors, we can continue to cherish those marvellous mechanisms of old.

References

1. Victor S. Clark, *History of Manufactures in the United States* (Washington: Smithsonian Institution, 1928), I, p. 2.

2. Michael C. Harrold, *American Watchmaking: A Technical History of the American Watch Industry, 1850-1930*, Supplement, NAWCC Bulletin, Spring 1984, p. 3.

3. Harry Chase Brearley, *Telling Time Through the Ages* (New York, 1919).

4. Charles W. Moore, *Timing a Century--History of Waltham Watch Company* (Cambridge: Harvard University Press, 1945), p. 12.

5. See Joseph A. Schumpeter, *Business Cycles* (1939) and his *Capitalism, Socialism and Democracy* (1942).

6. David S. Landes, "Watchmaking: A Case Study in Enterprise and Change," *Business History Review*, LIII (Spring 1979), 1-39.

7. Clark, *History of Manufactures*, p. 35.

8. U. S. Centennial Commission, *Reports and Awards*, VII, p. 117.

9. Clark, *History of Manufactures*, p. 365.

10. *Commercial and Financial Chronicle*, XXIV (Jan. 6, 1877), p. 3.

11. The history of these companies is treated in more detail in Part II.

12. M. Cutmore, *Watch Collector's Handbook* (1976), p. 47.

13. Clark, *History of Manufactures*, p. 161.

14. *The New York Times*, July 20, 1879, p. 10.

15. *The New York Times*, December 4, 1888, p. 5.

16. See the *United States v Keystone Watch Case Company*, 1911.

17. For the economics of the case, see W. B. Stevens, "A Group of Trusts and Combinations," *Quarterly Journal of Economics*, XXVII (August 1912), 602-08.

18. U. S. Department of Commerce, *Trade Information Bulletin*, No. 584, 1927, p. 1.

19. See the Report by Gilson G. Blake, Jr. on "Swiss Watch and Clock Industry," U. S. Department of Commerce, 1927, p. 7.

20. For a good history of the Swiss watch industry, see Landes, "Enterprise and Change," pp. 12-25.

21. This history is based on Anthony Gohl, "The Wrist Watch," NAWCC *Bulletin*, XIX (Dec. 1977), p. 587.

22. Gohl, "The Wrist Watch", p. 592.

23. Cutmore, *Watch Collector's Handbook*, p. 49.

24. Moore, *Timing a Century*, p. 211.

25. The best single source of these matters is U. S. Tariff Commission, *Concessions Granted by the United States in the Trade Agreement with Switzerland* (Washington, 1936).

26. August C. Bolino, *The Development of the American Economy* (Columbus: Charles E. Merrill, 1966), pp. 279-80.

27. Moore, *Timing a Century*, p. 277.

28. Harold G. Vatter, *The U. S. Economy in the 1950's* (New York: Norton, 1963), p. 81.

29. The references in this section may be found in August
    C. Bolino, "Who (or What Killed Waltham?" NAWCC *Bulletin*
    (October 1983), 555-557.

30. Harrold, *American Watchmaking: A Technical History,* p. 4.

PART I

The Beginnings

When America was first settled, European watches had three-wheel trains, verge escapements, and fusee drives. They were ornate, but rather poor timepieces. In the 17th century, several technical improvements were added: first, glass crystals, then springs, enameled dials and second hands. By the beginning of the 19th century, American watchmakers did not really make watches: they simply cleaned, repaired and fitted them. When they did make a watch, they finished it from a crude imported blank. The finished watch was indistinguishable from its English counterparts.

James Gibbs asked earlier, "Who was America's first Watchmaker?" (1) He states that no watch has been found that was produced in the colonies prior to 1711. The earliest American watchmaker was probably Thomas Harland (1735-1807) or Luther Goddard (1762-1842).

Harland was born in England, learned watchmaking there, and came to the Massachusetts colony in 1773. He advertised in December 1773 that he made "horizontal, repeating and plain gold watches in gold, silver, metal or covered cases." (2) But it is not certain that Harland actually made complete watches, because in a later advertisement, he did not claim to be a watchmaker but simply advertised "an assortment of warranted watches, vis. England silver watches, cap'd and jewelled, day of month and seconds, in silver and gilt cases; second-hand watch watches, various sorts; French gold and silver watches, day of month, seconds, and plain." This seems to remove him from contention as the first American watchmaker. However, his obituary in the *Connecticut Gazette* in April 1807 stated, "Died at Norwich, Mr. Thomas Harland, age seventy two, Goldsmith; he is said to have made the first watch ever manufactured in America." (3) But since no evidence has ever been uncovered to prove this statement, we probably should credit Luther Goddard with this distinction. However, Bailey declares, "It is probable that he was the first man to construct watches in America."

As we stated above, few watches were made in Colonial America. When Harland died in April 1807, he left 24 gold- and 93 silver-cased watches, but most were of English, Irish and French manufacture.

Luther Goddard

Luther Goddard (1762-1842) is not only better known than Harland, but he (Goddard) was the first American watchmaker to produce watches in quantity. Thus, he may lay claim to being the first American watchmaker. Goddard went to nearby Grafton in 1778 to learn clockmaking from his cousin, Simon Willard, but he returned to Shrewsbury to make ships and to farm. The Willard shop was established by the oldest of Goddard's first cousins , Benjamin, in 1764 at the family homestead, about 3 miles from the center of Grafton. Simon Willard, who taught Luther Goddard, and who learned his trade from his older brother, Benjamin, moved to Roxbury for the summer in 1780 and permanently in 1783. Since the Willard brothers moved constantly, it may be assumed that Goddard was self taught, as were most of the watchmakers of that time.

When the Jefferson embargo took effect, it cut off supplies from England. Goddard then converted to watchmaking. He enlisted his sons, Parley and Daniel, as apprentices and he hired addditional workers as his business warranted. Some of these were British soldiers, who knew watchmaking and who chose to stay in the United States when the Revolution ended. (4) In 1812, he made his first watch--of the verge type. We know from Mulvin that number 63 is an English-lever fusee, and that number 534 has a verge escapement with the large Eagle balance cock (the eagle was Goddard's trademark). (5) In all, he made about two watches per week, a total of about 600 to 650 watches, which were hand made.

Goddard made the plates, wheels, barrels, fusees, cocks and other brass parts, but he purchased imported dials, hands and springs from Boston. The movements were engraved with several variations of his and his sons' name: L. Goddard, L. Goddard & Co., Luther Goddard, L & P Goddard, and D. P. Goddard & Co. The cases were usually open-face silver and were made by Goddard or his helpers. According to Small, nearly all machines and tools in Goddard's shop were English . (6)

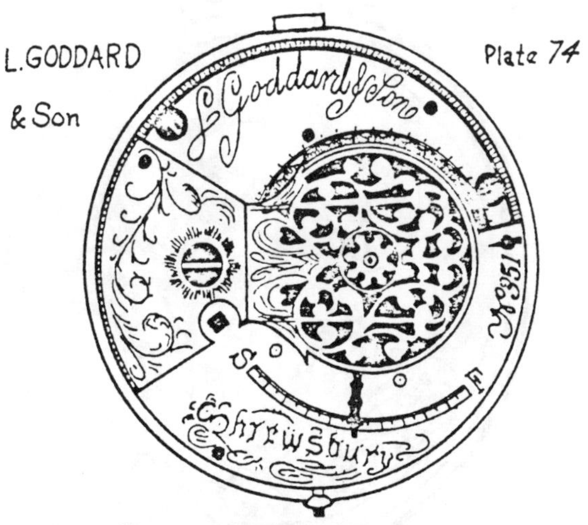

FIGURE 1 *L. Goddard & Sons Movement.   Sketch by Percy L. Small*

As long as the Embargo prevented English watches from entering the United States, Goddard's business flourished. His medium-sized, fusee-type watches were as good as similar watches produced in Europe, but when the Treaty of Ghent was signed on December 24, 1814, it removed the embargo on less expensive watches, thereby crippling Goddard's business. In 1817, Luther moved to Worcester with his son Daniel (Parley, who stayed in

Shrewsbury, joined them later).   Luther continued to make watches in Massachusetts until the death of his first wife in 1828, when he moved briefly to Connecticut.   He returned to Massachusetts, married again and remained active as a watchmaker until his death on May 25, 1842.

After Luther Goddard died, his sons continued the watch business for a while, but in 1870, Daniel Goddard built a shop for making clocks.   Some watches were produced in this clock shop, but the exact number is uncertain.

FIGURE 2   : *The H & JF Pitkin Movement.*
Sketch by George E. Townsend

The Pitkin Brothers

The next part of the story of the development of the Massachusetts watch industry takes us to the state of Connecticut, where James and Henry Pitkin, of East Hartford, produced machine-made watches in 1838. (7) The watches were marked "Henry Pitkin," "H. & J.F. Pitkin," and those produced in New York City, "Pitkin &

Co." They produced about 800 watches, of which "no more than three hundred were sold." (8) They made their own crude machines for making 16- and 18- sized watches, but the entire process was far too costly for the foreign competition of the time. Accordingly, they moved to New York City in the hopes of surviving there.

There were, in fact, four Pitkin brothers. Henry Pitkin was born on January 2, 1811 to Captain John Pitkin and the former Olive Forbes. He was the watchmaker and James Flagg (the youngest) was the business and financial manager. He had little interest in working with his hands; he preferred, instead, mathematics and bookkeeping. For this reason, he was the only one of the brothers not to serve an apprenticeship. His other brothers, John and Walter, had only a passing interest in watch manufacturing. Henry, however, was obsessed early with the idea of producing good quality watches using true mass production techniques with interchangeable parts.

Henry Pitkin, hearing of Luther Goddard's efforts in nearby Massachusetts, turned his own genius towards the design and construction of watchmaking machinery. Thomson claims that, "This machinery he ultimately designed and built." (9) Rosenberg goes further. He states, "These were the machines that later found their way to Roxbury and were models for more complex ones that were later in use at Nashua and Waltham. (10)

Since John and Walter Pitkin had been apprenticed to a silversmith, Jacob Sargeant, they built a shop in East Hartford to produce spoons, bowls and other silver items. When the business prospered, James (JF) joined them, and he convinced them to add a line of jewelry and watches. At this point, Henry became interested. He added a repair shop at the rear of the store and it was there that he trained several apprentices (including Nelson P. Stratton of later fame). By 1834, the design of machinery and watches was completed, and the first watch was produced in 1838--it was about 16 size, three-quarter plate, had a slow train (4 beats per second), and a going barrel.

All the watches produced at Hartford were sold at the company store.   Their biggest problem was finding cases.   There was an adequate supply of 18 size cases, but the Pitkin movements were smaller and thinner.   Henry found his answer in Boston, where he met Aaron Dennison.   Dennison referred him to John R. Proud, of the firm of Hiram and Daniel Tarbox, of New York City.   JF Pitkin eventually convinced Proud to move to Hartford to head the case department of the Pitkin Company.

Henry Pitkin, who Thomson calls the "Father of American Watchmaking," came to a very sad end.   The four brothers argued constantly, and the company moved to New York City as one consequence.   Henry had violent and "insane rages," and he committed suicide in 1846.   James sold the assets of the company to Aaron Dennison, Edward Howard, and Samuel Curtis, all of Boston. By this transfer, Henry Pitkin laid the groundwork for the mass production of American watches.

References

1. James Gibbs, "Who Was America's First Watchmaker?" NAWCC *Bulletin*, XVIII (December 1976), 556-565.

2. Chris Bailey, *Two Hundred Years of American Clocks and Watches* (Englewood Cliffs: Prentice-Hall, 1975), p. 65.

3. Bailey, *Two Hundred Years*, p. 191.

4. This history is taken from Percy L. Small, "Luther Goddard and His Watches," NAWCC *Bulletin*, No. 48 and No. 52 (April 1953), 355-363.

5. Walter J. Mulvin, "Two Sleepers," NAWCC *Bulletin*, VII (June 1956), p. 226.

6. Small, "Luther Goddard and his Watches."

7. Cooksey Shugart, *The Complete Guide to American Pocket Watches* (Cleveland: Overstreet, 1981), p.5.

8. Richard Thomson, *Antique American Clocks and Watches* (1968), 173.

9. Thomson, p. 174.

10. Charles Rosenberg, "Pitkin: Fact and Fiction," NAWCC *Bulletin*, X (Feb. 1963), p. 584.

CHAPTER 2:
EARLY MASS PRODUCTION
TECHNIQUES APPLIED TO WATCHES

The Concepts

Nearly everyone agrees that the success of the American watch industry in the nineteenth century was dependent on the application of the concept of interchangeable parts, but the precise way that this application developed is not fully known to watch collectors. Interchangeable parts were used in several industries before the idea was adopted in watchmaking. The process came to be known as the "American System of Manufactures." This system involved the light manufacturing of such durable goods as wooden-movement clocks, axes, typewriters, and watches. It is characterized by the mass production of interchangeable parts on specialized machinery arranged in sequential operation. In the United States, it was made possible by and encouraged by the growing population, which provided a large market.

According to Hoke, between the mid-1820's and the late 1840's, the federal armories shared the lead in innovations with the private sector, but by the 1860's and 1870's, the private sector took the lead, especially in watchmaking and typewriter manufacturing. (1)

The idea of producing similar (or identical) objects in large numbers is quite old. In modern times, we can look to the coinage as an early example. But even for coins, which had no moving parts, the job was difficult. The coins varied considerably in quality. Imagine the task of designing equipment to make products by the hundreds or thousands when the parts had to fit, mesh or rub against each other and where the tolerances had to be within a thousandth of an inch.

Interchangeable parts were first introduced at an exposition in Paris in the eighteenth century, but the idea lay dormant because the

French considered it an impractical scheme. However, in a letter written in 1785, Thomas Jefferson, then Minister to France, wrote the following after visiting a gunsmith named Le Blanc:

> An Improvement is made here in the construction of muskets, which it may be interesting to Congress to know, should they at any time propose to procure any. It consists in the making every part of them so exactly alike, that what belongs to any one, may be used for every other musket in the magazine. The government here has examined and approved the method, and is establishing a large manufactory for the purpose of putting it into execution. As yet, the inventor has only completed the lock of the musket on this plan. He will proceed immediately to have the barrel, stock, and other parts, executed in the same way. Supposing it might be useful to the United States, I went to the workman. He presented me the parts of fifty locks taken to pieces, and arranged in compartments. I put several together myself, taking pieces at hazard as they came to hand, and they fitted in the most perfect manner. The advantage of this when arms need repair are evident. He effects it by tools of his own contrivance, which, at the same time, abridge the work, so that he thinks he shall be able to furnish the musket two livres cheaper than the common price. (2)

Because the French judges usually focused narrowly on practical applications for determining prizes for inventions, they overlooked the exhibition of interchangeable parts, which appeared to be a curiosity. In 1806, the judges were even more practical in their approaches. They stated, "The products of a regular manufacture which flow into trade merit more praise than a tour de force, which often results only from the patience of one individual and imparts no development to the industry of a country." This approach seems to have produced some results, because as Hafter states, the French developed the Japy machine for making watches and Breguet's innovations--the second hand and the self-winding watch. (3)

The United States entered this industrial competition when the

War Department established the interchangeability of parts as a major goal, and it pursued this goal for nearly 50 years. In popular legend, Eli Whitney is the father of mass production. Whitney never quite achieved the stature that is attributed to him, but he foresaw the great potential of interchangeable parts, and he was successful in obtaining a government grant in 1798 from the War Department for the production of 10,000 muskets. He aimed "to make the same parts of different guns, as the locks, for example, as much like each other as the successive impressions of a copper- plate engraving." (4) This view has been challenged by Battison, who claims that Whitney did not produce interchangeable parts; his two-year contract took ten years to complete using standard techniques. (5)

The early successes came in areas where the tolerances were fairly large, ie in furniture and hardware. When the attempt turned to guns, clocks and finally watches, much more precision was required. But Landes adds another dimension--the cultural matrix. To quote him, "It was no accident that the watch industry found the answer it did to foreign industrial domination." (6)

The Breakthrough

The breakthrough came in clocks, because clock parts were much larger. Eli Terry (1772-1852), who learned to make clocks by the slow, conventional method, decided he could speed up production if he used a water-powered shop and specially-made machines. In 1806, he received a contract to construct 4,000 clock movements in 3 years. Terry spent the first year making machinery, to turn out interchangeable parts--probably the first such parts in United States economic history. He was able to produce 1,000 clocks in the second year of his contract and 3,000 in year three. Terry also became the grand clock teacher; many of his pupils became leading clock manufacturers, particularly Chauncey Jerome and Seth Thomas. As clock production became more sophisticated, the knowledge gained filtered down to the watch industry. But the first fully-integrated factory in the United States came in the textile industry, not in watches or clocks. (7)

Simeon North, a Connecticut Yankee, improved the methods of mass production. He was born on July 13, 1765 in Berlin, Connecticut, the fourth son of Jedediah and Sarah North. His manufacturing career started in 1795, when he began to produce scythes in an old mill adjoining his farm. On March 9, 1799, he obtained his first of several government contracts to produce 500 horse-pistols, but it was his contract of April 16, 1813 that is important here. It followed his recommendations and stated, "The component parts of pistols are to correspond so exactly that any limb or part of one Pistol may be fitted to any other Pistol of the twenty thousand" (8) North always met the standards imposed on him by contractual arrangements.

As a result of the work of Whitney and North, the United States was able to introduce a new "American System of Manufactures." The system was feasible because production was broken down into a number of small operations--what Adam Smith called in 1776, "division of labor." But before the system worked effectively, a number of technical barriers had to be overcome. These included:

```
        (1)  A device for exact measurement
        (2)  Machine tools and steel for
             these machines
        (3)  A supply of interchangeable parts
        (4)  Machine lubricants
        (5)  Inexpensive transportation for raw
             materials and for finished products
        (6)  Low cost power   (9)
```

By the middle of the nineteenth century, the railroad had reached to Chicago, the steel industry had learned how to puddle iron, the Brown and Sharpe Machine Company had perfected the vernier caliper, and American industries were converting to steam power.

Hounshell states that the American system of manufacturing has too often been "identified too exclusively with antebellum small-arms manufacture by and for the federal government." (10) For him, the manufacturing practices at the armories were a special case, and the most important characteristic was not interchangeable parts but rather the use of special tools for special purposes. Hounshell adds." If historians insist upon equating New England armory practice with the American System of Manufactures, they will misinterpret the technologies employed in the New England wood- and brass-clock industry." (11)

The American system was characterized by a number of operations, each performed by a different tool or machine, each requiring jigs, fixtures and guages to verify dimensions of parts. Terry's clock factory was fully-integrated and used other than hand power. The watch industry was integrated in the same sense. It performed many processing operations on a variety of materials to manufacture hair springs, main springs, brass parts and jewels. But many other American industries also integrated a number of processes under one roof.

Although the "armory practice" was introduced into other areas of manufacturing after 1850, there were some successful efforts in watchmaking before then. Cutmore tell us that, "From about 1840 onwards, machine-made watches with interchangeable parts were produced in Switzerland." (12) According to him. the machine for this operation was designed by Leschot at Vacheron & Constantin, the man who had improved Swiss lever escapements. But there was a major difference in European developments. Japy in France and the Prescott ebauche makers in Lancashire, England developed a number of hand tools. Production was geared toward finishing by hand rather than by machine, which was the American goal. In adddition, the Cutmore statement above seems to run counter to one made by Landes, who dated the breakthrough in the 1850's. As Landes said, this process could not penetrate watchmaking because "the smallness of watches and the degree of precision required entailed the application of far more exacting production standards than in woodworking and relatively rough metal work." (13)

In the 1850's, there were many industries using mass production techniques (watchmaking was only one of them). For example, by 1853, Singer Sewing Machine Company made 300 machines, each unique, but when the company hired William Perry as superintendent things changed quickly. He had been a machinist in the Colt Hartford Armory--a training ground for mass production. (14) Samuel Colt patented his first revolver in 1835 and by 1853, he built a large armory at Hartford, Connecticut. It contained 1,400 machine tools, many developed by his superintendent, Elisha K. Root. As one authority stated, "This plant probably represents the limit of achievement reached in the large-scale precision manufacture by the middle of the nineteenth century." (15) Root was one of a number of mechanics who, in the nineteenth century, made important innovations, which were crucial to the growth of mass production techniques.

By the middle of the century, the majority of machine tools now in use (except automatic lathes and gear cutting machines) already existed, but not in their current form. These tools, however, were incapable of producing small, precision parts. Thus, the early watch pioneers had to first develop the required tools, which was accomplished between 1850 and 1900. In this period, Americans invented or improved automatic feed mechanisms, control cams, automatic screw machines and the watch lathe. By the turn of the twentieth century, the American watch industry was a very technologically-advanced, manufacturing system.

Mass Producing Watches

The person who applied these mechanisms to watchmaking was Aaron L. Dennison, who in 1850 used a stamping machine and a form of upright lathe to cut watch plates. When the process succeeded, he built a 100 by 25 foot factory, which used steam power. However, the Pitkin brothers made the first attempt in the United States to mass produce watches in their Hartford plant. Although their design was not conventional, they had some success. The next part of he story, and the real beginning, was the meetings of 1848 and 1849 of Edward Howard and Aaron L. Dennison. Brearley describes these meetings as, "The contact of two minds was like the meeting of flint and steel." (16)

FIGURE 3   *Aaron L.Dennison and Edward Howard*

Edward Howard was born in Hingham, Massachusetts in 1813. At the age of 16, he was apprenticed to Aaron Willard Jr.--a leading clock maker. Dyer tells us that Howard was a mechanical genius and that "some of the clocks he made as a boy are as good to-day as when they were first put up." (17) Although Howard saw clocks being produced by machine, and he understood that the production of watches was controlled by the master craftsmen of Europe and ranked as a "fine art," it is not certain that he was interested in their production until he met Dennison. Howard understood that a variation of less than a thousandth of an inch could make a watch useless, and he wanted to invent automatic machines that would produce parts with such precision.

At the completion of his apprenticeship, Howard worked with a butter and egg business. He lasted less than a year, and in October 1834, he became an employee with Henry Plympton at 10 Theatre Alley in Boston, which produced the "Dearborn Patent Balance"--a precision scale. (18)

In 1840, Edward Howard started his own company. He built a small factory at 15 Hawley Street, Boston, Massachusetts. He was joined in this venture by David P. Davis--a fellow apprentice in the Willard Shop. In 1843, they were joined by Luther S. Stephenson, the brother-in-law of Henry Plympton, and Stephenson was added to the company name. This new firm had a salesroom at 42 Congress Street, Boston. It manufactured the Dearborn balances, clocks and regulators.

After living at several addresses in Boston, Howard moved to Roxbury in 1845, where he built a new factory at East and Prescott Streets. Howard continued to excel at producing balances, while he dreamed of designing watch-making machines. In fact, as early as 1840, he was appointed Deputy Sealer of Weights, Measures and Balances for the state of Massachusetts, and he was given a a contract later for the first postal scales used in the United States. (19) These scales were produced by the Howard and Davis Company (Stephenson had left the firm in 1847), which also manufactured fire engines. In this, too, Howard's genius for production was evident: the Howard and Davis Company was awarded a gold medal for its fire engines.

Aaron L. Dennison, the other link in the start of the Massachusetts watch industry, was born in Freeport, Maine in 1812.

(20) After early showing his mechanical bent, he was apprenticed at age 18 to James Cary, a clockmaker in Brunswick, for three years. But he left suddenly to go to Boston, where he volunteered his service to the firm of Currier and Trott. He didn't stay long, for he travelled much to New York City and back to Boston in 1839, this time to stay.

Dennison studied the economics of mass production. He was determined to obtain the necessary experience for producing quality watches. He offered to work free for Currier & Trott, and he was in their employ for one year. But then he started the first of many businesses--as a watch repairman. He "took a window for himself near the corner of Washington and Milk Streets." Dennison soon left and became associated with Tubal Hone, who had worked in Luther Goddard's shop in Shrewsbury, Massachusetts. The restless Mr. Dennison next went to New York City doing odd jobs and gaining information about watches.

In 1839, Dennison returned to Boston and established a new firm, Dennison, Adams & Co., at 67 Washington Street. In 1846, the company moved to 203 Washington Street and changed its name to A. L. Dennison & Co.--this time with Nathan Foster, a new partner.

As he continued his relentless search for watch information, Dennison ventured into several non-related enterprises. He was in retail jewelery; he was in silk farming in Brunswick, Maine; he was in the manufacture of boxes for the jewelry trade--which eventually became the Dennison Manufacturing Company--a highly successful corporation. (Dennison later established the Tremont Watch Company and the Dennison Watch case Company)

It seemed he was marking time, because he thought always of interchangeable parts. In 1845, he visited the Springfield Armory to study how he might model a factory. But all of this was a prelude for his linking up with Edward Howard. Although Dennison functioned on the same street as Howard, they did not meet until 1848. Then they met several times. Howard suggested that they collaborate in the production of fire engines, but Dennison convinced Howard that watches offered more promise of a return. (21) Samuel Curtis, of Boston, was their source of capital--he put up $20,000, and Davis, his partner in the scale business also contributed. They started in the Howard and Davis shop, but soon had to move to a new factory, which was completed in 1851. In this collaboration, Dennison provided the watch-making experience and Howard provided the mechanical skills to design the new equipment. Dennison went to England and Switzerland to study watchmaking. What he found disappointed him. As he said, he "found in them such workmanship as I should blush to have it supposed had passed from under my hands in our own lower grade of work." (22)

The new company, the American Horologe Company, had its own problems. It was proposed originally that the company produce 8-day watches, but through all these difficulties, the company continued to produce many things. As Small and Hackett said, "These speculative side lines were not always successful or profitable, and it is said that they lost $50,000 in an essay into the manufacture of sewing machines." (23)

In the spring of 1853, the first of the 30-hour watches was produced, stamped "Warren," and then the company name was changed to the Boston Watch Company. (24) At that time, there were 100 employees and the old financial problems arose again. Curtis had invested $20,000 earlier, as had Howard, and now funds were short again. This time capital came from New York City. Fellows and Schell, of Maiden Lane, pledged an additional 20,000, if they could be sales agents of the Boston Watch Company. A few watches carried the Fellows and Schell name.

The panic of 1857 brought the company into bankruptcy. In March, the company assigned machines, movements, parts and several watches to Charles E. Rice, who was a manufacturer of straw goods in Boston. The next month the firm was sold at auction to Royal E. Robbins and Tracy and Baker of Philadelphia for $56,000. Some of the tools and materials were taken by Howard and as he said,"I then returned to the first factory at Roxbury, when a new company was formed, Howard & Rice, using Rice's equipment and materials. Later a new company-- the Howard Watch and Clock Company, was organized with a nominal capital of $150,000." (25) Dennison stayed on as superintendent of the new company of Tracy, Baker and Company in Waltham until 1861. When Baker sold his interest, the company became Appleton Tracy & Company. After other name changes, this company became the Waltham Watch Company.

Conclusion

In reviewing the efforts of these watch pioneers in Massachusetts, how can we conclude? Howard, himself, says, "When I look back and bring to mind what I went through physically and mentally to start and perfect the watch business, I am astonished at the endurance and perseverance with which I stuck to the task, and my friends are more astonished than myself." (26) But William H. Keith put things in better view, "Both of these gentlemen (Dennison and Howard) are certainly worthy of the highest commendation, in that they, assisted or unassisted, put foreward measures to accomplish, with or without the principles and practices being known to them, which had been employed by Eli Whitney, and subsequently produced at the Arsenal at Springfield, Mass., and by the clockmakers of Connecticut, and applied the same to the manufacture of watches, and produced goods that were a credit to themselves and acceptable to the public and the trade." (27)

Someone once said that imitation is the best form of flattery. If so, then Dennison and Howard must have been proud of their technological accomplishments, because the English and Swiss came eventually to use the American system of watchmaking. According to Aked, E. W. Streeter built "The first English watch manufactory." Aked quotes from the *London Times* of December 1868: "Mr. Streeter, of 37 Conduit Street, Bond Street, has followed the example of the Americans, in producing his watches by machinery. whereby a saving in cost of one-third is effected." (28)

The conclusion of Donald Hoke's doctoral dissertation is also relevant here. He sees the development of the American system of manufactures as an interaction between the mechanic and the entrepreneur. They usually knew each well, and their working relationship resulted in more investment in improved technology.

There is another conclusion that is equally important: The history of watchmaking is part of a much bigger story. While Waltham and Elgin were mass producing watches using specialized tools and machines, the American System of Manufactures was not feasible in all industries in the nineteenth century. The McCormick Reaper Company, for example, produced thousands of reapers, each fitted by hand. Even today, there is a similar dual economy. (29) Averitt distinguishes small batch and large batch production, and he equates the latter with mass production and the assembly line. The key to the success of mass production, according to him, was long-term planning and production schedules that were not tied directly to orders or sales. By comparison, small batch production was geared to customers requirements. This type of production was prevalent in the clothing, furniture and machine tool industries. Seen in this light, the early watch producers were bold innovators who took substantial risks and who received good returns when they succeeded.

References

1. Donald Hoke, "Ingenious Yankees, The Rise of the American System of Manufactures in the Private Sector," *Journal of Economic History*, XLVI (June 1986), p. 489.

2. *A History of Technology*, IV, The Industrial Revolution (Oxford: At the Clarendon Press, 1958), 437-438.

3. Daryl M. Hafter, "The Business of Invention in the Paris Industrial Exposition of 1806," *Business History Review*, LVIII (Autumn 1984), 322.

4. *A History of Technology*, IV, p. 438.

5. E. Battison. *Muskets to Mass Production* (American Precision Museum, 1976).

6. David S. Landes, *Revolution in Time* (Cambridge: Belknap Press, 1983).

7. S. Dunhill, *The Run of the Mill* (David Godine, 1978).

8. S. N. D. North and Ralph North, *Simeon North, First Official Pistol Maker of the United States* (1913).

9. August C. Bolino, *The Development of the American Economy* (Columbus: Charles E. Merrill, 1966), p. 76.

10. David A. Hounshell, "The System: Theory and Practice," in Otto Mayr and Robert C. Post (eds.), *Yankee Enterprise* (Washington: Smithsonian Institution Press. 1981). p. 127.

11. Hounshell, "The System," p. 128.

12. M. Cutmore, *The Watch Collector's Handbook* (Rutland: Charles E. Tuttle, 1976), p. 42.

13. David S. Landes, "Watchmaking: A Case Study in Enterprise and Change," *Business History Review*, LIII (Spring 1979). 26.

14. Joseph W. Roe, "The Colt Armory," in *English and American Tool Builders* (New Haven, 1916).

15. *History of Technology*, IV. P. 439.

16. Harry C. Brearley, *Telling Time Through the Ages* (New York: 1919), p. 168.

17. George L. Dyer, *The Story of Edward Howard* (Boston: E. Howard Watch Works, 1910), p. 6.

18. Percy L. Small and F. Earl Hackett, "E. Howard: The Man, The Company," NAWCC *Bulletin*, Supplement, 1962, p. 2.

19. Small and Hackett, pp. 3-4.

20. This brief history is from Charles S. Crossman, *The Complete History of Watchmaking in America*, 1885-1887.

21. Orra L. Stone, *History of Massachusetts Industries* (Boston, 1930), p. 944.

22. Brearley, *Telling Time*, p. 168.

23. Small and Hackett, "E. Howard," p. 6.

24. Stone, *History of Massachusetts Industries*, p. 945.

25. E. Howard, "American Watches and Clocks," in Chauncey M. Depew (ed.), *One Hundred Years of American Commerce* (New York: D. O. Haynes and Co, 1895), p. 542.

26. Crossman, *Complete History*, p. 26.

27. Crossman, p. 27.

28. Aked, Charles K. "The First English Watch Manufactory," NAWCC *Bulletin*, XXVIII (Feb. 1986), 52-60.

29. Robert Averitt, *The Dual Economy* (New York: Norton, 1968).

PART II

Individual Companies

Introduction

In this part, we present a brief history of the Massachusetts companies that produced watches. But how shall we classify them? Chronologically? This would introduce the problem of conflicting and overlapping dates. Or should we list the companies by number of watches produced. Waltham, obviously, would be first with 35 million. Then would come Hampden with 8.2 million (including 20,000 as the New York Watch Company), The United States Watch Company with 802 thousand and Howard with a total of 110 thousand movements. But after the big four we are in trouble again, because the output of such companies as Tremont, Auburndale and Fitchburg is a matter of dispute. In fact, Fitchburg produced only one watch. Or we could list the companies by quality of output (what a subjective idea!) Would Howard be first or Waltham? Alas, we solve the problem simply: we take up the companies alphabetically.

Meager Beginnings

As we stated in Part I, the idea of a machine-made watch with interchangeable parts was pursued by several persons in the mid-nineteenth century. A partial success was achieved at the Waltham Watch Company, but there was a large potential market for an inexpensive watch. Those who labored after this goal were convinced that a radically-new design was necessary to achieve success.

The primary obstacle in manufacturing an inexpensive watch was the temperature error and what Battison called "poise." (1) Lack of poise caused the watch to run in an erratic fashion, because all parts were not synchronized and there was "position error."

The Breguet tourbillon was one solution for correcting these design errors. His design mounted the escapement in a frame that

Table 1

AUBURNDALE WATCH COMPANY CHRONOLOGY

Date	Item
1875 (July 20)	Jason R. Hopkins registers patent No. 161513
1876	William B. Fowle invests in the Hopkins rotary watch
1877	First Auburndale movements produced
1878 (May 28)	W. A. Wales patents a chronograph ("The Auburndale Timer")
1879	The Auburndale Watch Company in corporated
1881	Production of timers ceases
1883	Auburndale Watch Company fails
1884 (Feb.)	All machinery and parts sold

revolved one time per minute. This produced little error in the escapement, but the precision for mass-producing the frame did not exist at that time (about 1800).

From the concept of the revolving escapement came the natural followup of revolving the entire movement. This would remove the problem of the precision frame, mentioned above. In an early version of this idea, Bonniksen produced a "Karrusel" watch that rotated in 52.5 minutes. (2)

Enter Hopkins and Fowle

On July 20, 1875, Jason R. Hopkins registered patent number 161,513 for a watch patterned after the Karrusel. (3) It had a very large mainspring to drive the rotating movement (the mainspring was, in fact, two springs riveted together), which turned once in 20 minutes. The balance, which contained the only jewels in the watch, turned on its own axis, as in the tourbillon. Hopkins, a native of Maine, who lived and died in Washington, D. C., was a watchmaker, but he also produced pianos and clock cases. (4) When William D. Colt, an attorney from Washington, D. C., heard of the Hopkins watch, he (Colt) arranged a meeting between Hopkins and Edward A. Locke, of Boston, who was interested in producing a low-priced watch. After the meeting, Hopkins improved his design and offered it to Benedict and Burnham of Waterbury, Connecticut. They rejected the design. Hopkins then offered his watch to George Merritt, also of Boston, who supplied some funds for additional design developments, but he also withdrew from the project. (5)

In 1876, Hopkins and W. A. Wales visited Auburndale, Massachusetts to propose to William B. Fowle that he support financially the production of the Hopkins rotary watch. Fowle, who resided in Auburndale, was told that with an investment of but $16,000, the company could produce 200 watches per day. (6)

William B. Fowle was born in Boston on July 27, 1826. He worked as a ticket master on the Boston and Worcester Railroad in 1848, and until the Civil War, he was a merchant of sorts. In that war, he was Captain of the 43rd Massachusetts Volunteers. When he was mustered out of service, he returned to Auburndale and commenced an association in the coal business, which lasted until he began his considerable investment in the rotary watch, which according to Crossman totalled $140,000.

When the decision to produce watches was reached, a factory was built in Weston, across the Charles River, near Fowle's home. The plant was 2-story, measuring 40 feet by 20 feet and 32 feet by 20 feet. Fowle reached the factory by private ferry. The machinery for the new plant was obtained from the United States Watch Company at Marion, New Jersey, which ended production in 1874. The bulk of the equipment came from George E. Hart, who was disposing of the company's assets.

The first movements were produced in 1877. They were 18 size, lever escapement, stem winding, stem set, with open- faced cases. The trains and bridges were placed on top of the very large mainspring barrel. The price was $10 to the trade, but considering the poor performance of the watch, it was overpriced. Many watches were returned to the company for repairs or refund. The idea for the rotary was sound, but the technology was wanting. After producing about one thousand watches, the rotary was abandoned. (7)

The production of an inexpensive watch fared better elsewhere. In 1877, Daniel A. Buck developed a watch for Merritt and Locke, who had rejected the Hopkins rotary. Buck was a master machinist with an inventive flair. (He produced a miniature engine for exhibition) His first model was a rotary with a long wind, but his next model was a completely new layout with a duplex escapement. It was a relatively simple watch, with fewer parts, and the plates could be stamped out of sheet brass. It sold for $3.50 in 1878, although the price was reduced to $3.00 when production accelerated.

Although Buck's watch was accepted in the marketplace, it had some technical deficiencies. The long wind (140 half turns) was often a nuisance, and the hands were set by pushing them with the fingers to the right time. As Olson wrote, "True, the Waterbury Watch was a success and the Auburndale Rotary a failure, but I believe that if an equal amount of money had been expended in development of this watch and the machinery for producing it as on the Waterbury, the Auburndale would have come to a more fitting end." (8)

Although the rotary failed, the Auburndale Watch Company continued to produce watches. Fowle, because of his large investment in the company, pushed for new models. On May 28, 1878, W. A. Wales patented a chronograph. It came to be known as the "Auburndale Timer." It was constructed so as to record 1/2, 1/8th and 1/10th seconds. The original model was side stop, but later models were stop and fly back at the stem. Several persons improved the original. A. Craig, the master mechanic, invented the stop and start mechanism, and J. H. Gerry the new escapement. The cases were made by The Thiery Watch Case Company, of Boston, which also cased the rotary. (9)

The timers were reliable and the public responded favorably. They were purchased in the United States and overseas. But it is obvious that a watch company would have a difficult time surviving on the production of just a timer. Its demand was too erratic and too seasonal (for example, for use at races). At times, sales reached 400 per month, but often slumped to near zero. It was clear that the company needed additional products to level out its seasonality of sales. (10)

The first superintendent was Warren E. Ray, a neighbor of

FIGURE 4   *The Auburndale Timer, Number 1836.   Note patent date, May 28, 1878, near balance wheel*

Fowle, but he died suddenly in October 1876, and he was succeeded by James H. Gerry, who was associated earlier with the Waltham and United States Watch Companies.   The Auburndale Watch Company suffered the same employee turnover problems of other new companies.   Gerry left as soon as the Timer was in production.   The superintendent's job had a succession of persons.   After Ray died, he was succeeded by William Guest, who was succeeded by Gerry in 1877 and by George H. Bourne, E. H. Perry, J. Hinds and O. L. Strout. (11) In addition to the above, Battison's compilation includes some famous and not so famous employees: L. C. Brown, Frederick H. Eaves, Jose Guinan, Sadie Hewes, Isaac Kilduff, E. Moebus, James O'Connell, Frank N. Robbins, John Rose, Thomas W. Shephard, William H. A. Simmons, Alfred Simmons, Thomas Steele and George Wood. (12)

In 1879, the company was incorporated, with an authorized capital of $500,000, and with William B. Fowle as President, George

H. Bourne as Secretary-Treasurer and Chauncey Hartwell as Manager. Hartwell was a watchmaker with experience at Waltham Watch. He developed an inexpensive 18 size, 3/4 plate watch to fit regular hunting cases. The watch was sold in two grades: the "Bentley" and the "Lincoln"--named after Fowle's sons. The Bentley was stem wind; the Lincoln key wind. This attempt was also an economic failure, and the production was discontinued. Instead, the company turned to making metallic thermometers, along with the timers. The thermometers were pocket size to 18 inches in diameter, with nickel cases. Some were made for medical and scientific uses.

The production of these items lasted until 1881, when manufacture of the timers ceased. Technically, the watch was a success, but it was managed badly. Advertising was generally lacking and sales fell in the off season. The income from thermometers could not sustain the entire operation, and the company failed in 1883. In February 1884, the machinery was sold.

The final question is , how do we rank Auburndale in horological history? The answer is that it stood between the near success of the Pitkin Brothers in 1840 and the successful dollar watches that flowed from the Waterbury Watch Company and those that followed it. The Pitkins, Hopkins and Buck were the pioneers who established the technical basis for producing inexpensive watches. From their efforts over 40 years, we arrived at the proper procedures: the manufacture of inexpensive watches, of brass, with no jewels, few parts and a very inexpensive dial and case.

References

1. Edwin A. Battison, "The Auburndale Watch Company," Musuem of History and Technology, *Bulletin 218* (1959).

2. Paul M. Chamberlain, *Its About Time* (New York, 1947), p.229.

3. In fact, Hopkins was granted two patents on the same day for a watch that he hoped would sell for 50 cents.

4. Battison p. 53.

5. Charles S. Crossman, *The Complete History of Watchmaking in America*, Jewellers Circular and Horological Review, 1885-87 (Adams Brown reprint), pp. 400-401.

6. Crossman, p. 149.

7. Michael C. Harrold, *American Watchmaking: A Technical History of the American Watch Industry, 1850-1930*, NAWCC *Bulletin*, Supplement, Spring 1984), p.55.

8. David C. Olson, "The Auburndale Rotary," NAWCC *Bulletin*, VIII (April 1959), p.510.

9. Crossman, p.148.

10. Battison, p.64.

11. Henry G. Abbott, *The Watch Companies of America* (Chicago: G. K. Hazlitt, 1888), p. 94.

12. Battison, p.60.

CHAPTER 4:
THE FITCHBURG WATCH COMPANY

It is best to begin this chapter by quoting Crossman. "The Fitchburg Watch Company never really existed. although all the necessary preparations had been made in the expectation of forming a watch company in Fitchburg Mass." (1) Is Crossman correct? Was there a Fitchburg Watch Company? Did it produce any watches?

Getting Started

Sylvanus Sawyer is usually credited with being the founder of the "Watch Manufactory" at Fitchburg. He was a stockholder in the U.S. Watch Company, of Marion, New Jersey. which according to Shugart, produced 286,000 watches from 1864- 1872. (2) It got caught in the great depression of 1873; it lowered prices on most grades of its 18 size movements. but this could not save the company. In 1874. it sold its machinery to the Fredonia Watch Company and the newly-formed Fitchburg company. Sawyer. a native of Fitchburg. decided to start a watch company in his home town. at first by producing tools and machinery. In 1875. he hired Henry J. Lowe. who had been superintendent of the U. S. Watch Company at Marion. Lowe worked for over a year, but poor health forced him to retire, and he was replaced by Gilmore Crowell. a mechanical engineer with 19 years of experience.

According to Crossman. the list of employees included: Charles Whitehouse, William Guest. A. R. Bardeen. C. Vanderhoff. Charles Dodge and Charles Parker. This list varies from the one given in the *Daily Sentinel*:

Crossman	Sentinel
Sylvanus Sawyer	Sylvanus Sawyer
William Guest	William Guest
Thomas Parker	Thomas Palmer
Gilbert Crowell	
Charles Whitehouse	
A. R. Bardeen	
C. Vanderhoff	

Table 2

FITCHBURG WATCH COMPANY CHRONOLOGY

Date	Item
1874	Sylvanus Sawyer founds a "Watch Manufactory" at Fitchburg, MA
1875	The Farwell Place -- a two story building with engine room is offered to the company by the Fitchburg Board of Trade
1875 (June)	The Fitchburg Board of Trade commends the Watch Company for the quality of its watch
1876	The Fitchburg Watch Company lists five employees
1878	Part of the machinery sold to the Cornell Watch Company
1880	The Sawyer Watch Tool Company begins operation in Fitchburg
1880 (July)	Company receives its first order for machines
1880 (Dec.)	Company receives highest award for watch machinery at 13th annual exhibition of the Massachusetts Charitable Mechanics Association
1885	Company producing watch lathes under Kesselmeier's patent

Charles Dodge

M. A. Goodrich
George D. Colony
Michael Kivlon
John Haskins
Fred Fosdick
Henry Piper

The *Sentinel* stated that the above-named persons had adopted articles of association for the purpose of forming a corporation for the manufacture of watches. (3) The company was to start with 500 shares of $100 each, with the privilege of increasing the capital to $150,000.

On June 17, 1875, the Board of Trade created a Special Committee on the Watch Manufactory. It reported that "the watch movement, made under the direction of Mr. Henry J. Lowe, are (sic) of superior quality and workmanship." (4) It suggested that the factory start by making five watches per day and increase to twenty "as the demand increases." It recommended that since the company would need several years to grow that it lease the buildings instead of "putting a large amount of capital into the same." The property in Farwell Place, owned by Mssrs L. M. and E. T. Miles. was offered for this purpose. It was 60 feet by 40 feet, with two stories and an attic and an engine room. The engine was a Campbell and Whittier-- one cylinder of 6 horsepower. The Committee suggested that another could be added at a small expense." As an inducement to begin making watches, the building was offered "for two years without rent, taxes or insurance." After two years, the building was to be leased at $600 per year, and an additional five years at $800 per year. (5) The Committee concluded as follows, "Your Committee believe that the watch manufacturing business can be started here upon plans substantially as herein recommended. which should be classified among the best manufacturing concerns in our place." The report was signed by Rodney Wallace, Eugene T. Miles and H. F. Coggshall.

In 1876, *the Fitchburg Directory* listed five employees of the

"Watch Factory:" (6)

> Lowe, Henry J.
> Supt., watch factory
>
> Crowell, Gilmer
> emp. watch factory
>
> Whitehouse, Charles H.
> emp. watch factory
>
> Marshall, Charles
> emp. watch factory boards
>
> Vanderhoff, C. F.
> emp. watch factory boards

It is clear that these persons were not producing watches, because two years later the Board of Trade was still trying to interest investors in a company to produce watches. This quote from the *Daily Sentinel* tells the story: (7)

> While it is generally admitted that the establishment of the manufacture of watches in our city would seem to be a highly desirable consummation for the interests of the municipality in general, and while the promise held out to investors, if not immediately remunerative would at some distant day be likely to be a very satisfactory operation, there appears to be, not withstanding, a positive disinclination among monied and business men at this time, to touch anything of this nature, largely on the principle that a burned child dreads the fire.

Sawyer's search for funds was unsuccessful, and since his own $45,000 had run out, he ceased operations until some investors could be found. Part of the machinery was sold to the Cornell Watch Company and part was retained to be used later.

The Sawyer Watch Tool Company

In 1880, the *Daily Sentinel* stated that "The Sawyer Watch tool Company is a promising manufacturing enterprise recently commenced in this city." (8) This business grew out of Sawyer's attempt in 1875 to establish a watch factory in Fitchburg. A stock company was created, and when part of the stock was subscribed, he commenced to manufacture machinery. He had some success with "watch, clock, chronometer and jewelry machinery." He received his first order for machines in July 1880, when he employed several apprentices to work with his skilled craftsmen.

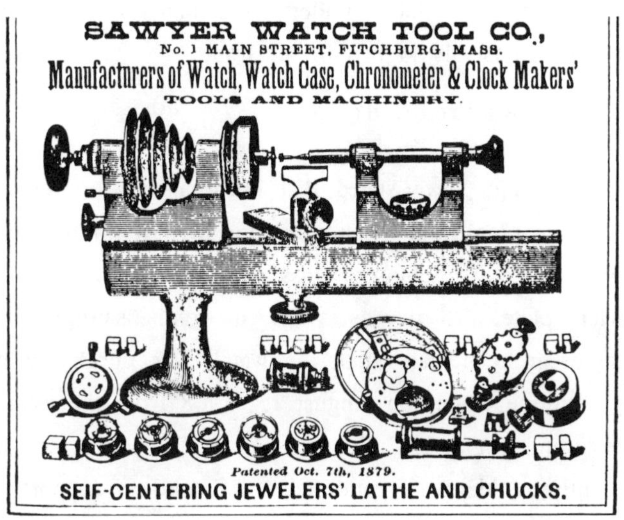

FIGURE 5   *The Sawyer Self-Centering Lathe*

In December 1880, the plant employed 12 employees, but had enough orders to employ 25 persons. Sawyer was considering a plan "to seek larger accomodations in the spring." The company produced five sizes of lathes. A new self-centering lathe was patented on October 7, 1879--a valuable improvement for watch makers. The company also produced screw-cutting machinery, a pinion-cutting engine, an automatic leaf polisher, and an automatic staff and pivot-turning machine. In all, it produced "75 different patterns of tools and fixtures."

Sawyer received the highest award for watch machinery at the 13th annual exhibition of the Massachusetts Charitable Mechanics Association. The company had received orders for equipment from

Holler Chronometer Company, the Auburndale Chronograph Company, Hall's Safe and Lock Company, the E. M. Flint's Safe Lock Company and Robbins and Appleton.

If Crossman is correct, then Sawyer continued to manufacture watch machinery until 1885, because as Crossman stated, "He at present manufactures a lathe under Kesselmeier's patent of Galion, O. but is not extensively known in the trade."

Were Any Watches Produced?

Several writers have stated that no watches were ever produced at the Fitchburg plant, but Selchow claims to have found a watch with "Union Watch Co., Fitchburg Mass" on the plate. It is 17 size, three-quarter plate, 15 jewels, with a compensating balance. (9) Selchow's research indicates that the watch was purchased by a New Hampshire antique dealer on August 15, 1968 from the estate of Clara Lowe, the granddaughter of Henry J. Lowe. Selchow concludes that, "the substantive evidence as outlined above tend to support the theory that this is, in all probability, a watch which was produced at Fitchburg, to be used as a model." (10) He states that the name Union Watch Company was the real name of the watch company Sawyer was trying to establish, and that the name 'Fitchburg' referred only to the place of origin. As we stated above, the Board of Trade commended Sawyer for the quality of his watch at a meeting in June 1875. This is undoubtedly the watch to which Selchow refers. No other watch has ever been found that was produced at Fitchburg.

References

1. Charles S. Crossman, *The Complete History of Watchmaking in America* (Adams Brown Reprint, Jewelers Circular and Horological Review, 1885-87), p. 158.

2. Cooksey Shugart, *The Complete Guide to American Pocket Watches* (Cleveland: Overstreet, 1981), p. 241.

3. *Fitchburg Daily Sentinel*, Oct. 8, 1875, p. 3. It should be noted that Crossman's Thos Parker is actually Thomas Palmer (a mistake repeated later by Frederick M. Selchow, "The Watch Company of Fitchburg Massachusetts," NAWCC *Bulletin*, April 1969, p. 914).

4. *Fitchburg Daily Sentinel*, July 13, 1875, p. 2.

5. E. T. Miles, mentioned as one of the owners of the building to be leased, was a member of the Special Committee on the Watch Manufactory. Apparently, in those days noone worried much about conflicts of interest.

6. Selchow, "The Watch Company," p. 917.

7. *Fitchburg Daily Sentinel*, May 14, 1878, p. 3.

8. *Fitchburg Daily Sentinel*, December 21, 1880, p. 2.

9. Selchow, "The Watch Company," p. 917.

10. Selchow, p. 920.

CHAPTER 5:
THE HAMPDEN WATCH COMPANY OF
SPRINGFIELD, MASSACHUSETTS

The Early Companies

The Hampden Watch Company was a result of a number of interrelated moves and consolidations, which was a pattern of early American watch companies. In the case of Hampden, the success of the company was due to Don J. Mozart, who was born in 1820 in Italy. His father, who was a watchmaker, moved to Boston in 1823. According to Crossman, young Don was taken as a very young boy aboard a sailing vessel, which went on a three-year cruise. (1) When he returned to Boston, he could not locate his parents, and he just drifted from place to place in search of them, while he worked at various jobs. He must have inherited mechanical skills from his father, so it was natural for him to follow in the watch trade.

Mozart was married in 1854 and settled in Xenia, Ohio until 1863, when he moved to Bristol, Connecticut to manufacture clocks and to put some of his own mechanical genius to work. In fact, he was granted a patent (number 25,034) for a dead-beat clock escapement. He thought much of starting a watch company, but he needed capital. In 1864, with the help of George Samuel Rice, of New York City, he formed the New York Watch Company. The capital totaled $100,000 and the main office was located at 180 Broadway. The investors included E. E. Haywood and J. A. Briggs-- the leading jewelry house in New York City, which manufactured rolled gold plate; John T. Mauran, Henry F. Fuller, L. T. Best, J. A. Simmons and Joseph Anning. George Rice was elected President and A. J. Briggs Secretary and Treasurer. Mozart served as the Superintendent.

The factory was located in Providence, Rhode Island, because some of the investors had business connections there and because they assured Mozart that there was a good supply of machinists in that

HAMPDEN WATCH COMPANY CHRONOLOGY

Date	Item
1864	Don Mozart and George S. Rice form the New York Watch Company
1864	Factory opens in Providence, RI
1867	Company buys old machine works in Springfield, MA
1870 (Apr. 23)	Fire destroys factory building; Company moves to two-story boarding house
1870 (July)	Production resumes on 18-size, three-quarter plate movements
1870	Company Charter moved from New York City to Springfield, MA
1875	Company closes because of shortage of funds
1875	A new company formed: the New York Manufacturing Company
1876	The New York Manufacturing Company fails
1877 (Jan.)	The Hampden Watch Company formed from old company
1877 (June)	Movements remodeled and factory opens again
1881	A new brick building completed. Five buildings used; Production at 400 watches per day
1886	Company sold to the Dueber Watch Case and Manufacturing Company
1888	Many employees of the Hampden Watch Company move to Canton, OH

city.   He set up shop on Broad and Weybosset Streets making machinery and tools.

Mozart had two ideas that would sustain him: he imported Swiss machines and remodeled them for his own purposes, and he pursued an idea of making a three-wheel watch. (2) He wanted to mesh wheels with a large number of teeth to pinions with a small number. Crossman tells us that although Mozart thought his "invention" was new, that this type of watch had been produced in Europe earlier. He states that he repaired and sold one that was of English make. (3) The remainder of Mozart's watch was more usual, except for the escapement, which he called a chrono-lever.   He believed that he could make an escapement so perfect that it would be free of friction and hence require no jewels.   In addition, he thought that his new escapement would lead to economies of scale and large profits.

George Samuel Rice provided the principal financial support. He encouraged the Hayward & Briggs jewelry firm to invest in the new firm.   With 20 employees, the company turned out some of the unusual 18 size chrono-levers.   However, Mozart's venture was not a success, and by mutual consent, he left the company and went to Ann Arbor, Michigan, where he started the Mozart Watch Company.   It too failed in 1870.   After Mozart's departure, the New York Watch Company hired L. W. Cushing, of Waltham, Massachusetts to make tools and machinery to produce a regular watch.   All the Mozart tools and equipment were discarded.   At this time, the company was offered the old machine works in Springfield, Massachusetts for only $37,500, payable in stock of the company.   In effect, the company took in more investors.   To accomodate them, the company was reorganized and the total capital was increased to $500,000.   The old stockholders were paid $30,000 for the tools and equipment that were made in Providence.

Moving to Springfield

The new plant was located four or five blocks from the Springfield Armory at State and Federal Streets. When the company was moved in October 1867, it had the following officers: George Samuel Rice, President; J. A. Briggs, Secretary and George Walker, Treasurer. The post of treasurer changed often. George L. King succeeded Walker, Theo L. Studley succeeded King, and finally Homer Foote took over the job in 1870. The new plant was located on Armory Hill--one mile north of the center of Springfield. The superintendent and foremen included:

```
O.P. Rice            Business manager
J. H. Gerry          Superintendent
George Griffin       Foreman, plate room
E. P. Gerry          Foreman, escapement
                        room
H. J. Cain           Foreman. balance room
Leonidas Murray      Foreman, train room
B. Gerry             Foreman, flat steel,
                        stem winding room
George Pollar        Foreman, gilding room
Charles Avery        Foreman, dial room
Charles P. Crafts    Foreman, machine shop
```

The company proposed to manufacture two grades of 18 size movements to be engraved "Springfield" on the plates, but before too many were made a fire destroyed the factory building on April 25, 1870. Fortunately, the company had received a two-story boarding house as part of the original purchase of the machine works, so it moved to this location and a new story was added. Most of the machinery and parts were not affected by the fire.

In July 1870, watch production resumed. Several grades of 18 size, three-quarter plate movements were produced, with the "George Walker" line being the highest grade at $200. The company also commenced to make a 16 size watch, with the name "New York Watch Company" on the dial. The lowest grade 18 size watch was the "John Hancock," which had his signature on the plates; the highest grade was the 19 jewel "Springfield." (See Table 4 for production statistics for 1872 to 1874).

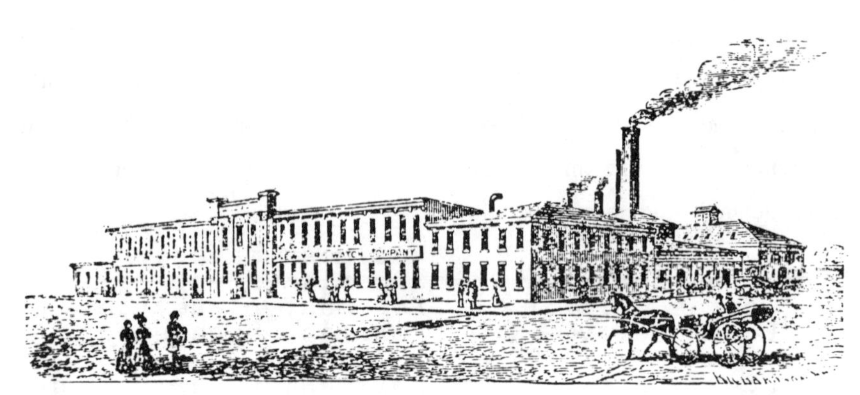

FIGURE 6   *The New York Watch Company Factory, April 1870.   Taken from Henry G. Abbott, The Watch Factories of America*

After the fire, the charter was moved from New York City to Springfield, Massachusetts, and several administrative changes were completed.   J. H. Gerry retired as Superintendent, and he was succeeded by Osmore Jenkins; John C. Perry succeeded O. P. Rice as General Manager; Henry J. Cain was appointed manufacturing Superintendent; and Homer Foote, who was elected Treasurer a short time earlier, was made Financial Manager.   By this time, $287,000 of the $500,000 authorized capital was realized.

Misfortune struck the company in 1873.   O. P. Rice resigned as President, amd his successor died in March after only two months in office.   Aaron Bagg was chosen to fill the vacancy, but he was forced to cope with the greater financial panic of that year, that was to last until 1880.

In 1875, the company was forced to close because of the extreme shortage of funds, but it was not idle for long, as a new

Table 4

THE NEW YORK WATCH COMPANY PRODUCTION STATUS[*]
(1872 – 1874)

Model	1872	1873	1874	Total
Woolworth	607	2,293	1,859	4,759
Studley	903	1,129	727	2,759
Haywood	16	1,417	694	2,127
Bagg	22	505	559	1,086
Norton	281	199	151	631
Rice	16	528	44	588
Foot	256	38	21	315
Billings	17	61	223	301
Hancock	15	165	104	284
Clark	27	32	42	101
King	2	4	33	39
Perry	0	0	24	24
NY Watch Co.	5	1	1	7
Briggs	2	0	0	2
Guy	0	1	0	1
			Total	13,024

[*]Key wind only

Source: R.J. Ziebell, Serial List, New York Watch Company (Ipswich: Old Post Office Clock Shop, 1972).

organization arose from the closure: The New York Manufacturing Company. This venture also failed, and the doors were closed again in 1876. But the company that arose from these ashes was a success. In January 1877, the bondholders, who held $300,000 in securities, bought the company; they named it The Hampden Watch Company. A new board of directors was formed: Homer Foote was chosen President, H. J. Cain Superintendent and Charles D. Rood came from New York City to become Treasurer. James D. Bauer and A. Breever also were selected to serve on the board. (4)

The movements of the old company were remodeled and the factory opened in June 1877. Gibbs describes it in great detail. (5) The dimensions were 30 feet by 120 feet (see Figure 7) It was operated by an engine of 85-100 horsepower and two boilers of about 100 horsepower. The windows gave ample light for manufacturing. It had 14 separate departments, and all machinery was made for the purpose. Each watch was kept in the adjusting room until accurately timed.

The new company was an immediate success. The *Fitchburg Sentinel* reported that in January 1878 the company received an order for 1,500 watches and this was followed by a second order for 1,000, mostly for the the South American trade. (6) At that time, about 50 watches were produced per day by a work force of 90-100 persons. The company was making 30 varieties of movements. In February 1878, it marketed its first "State Street, Springfield, Mass." watch--a solid nickel, stem winder, and it was working on its new chronograph or "fly-back" stop watch. To stimulate sales, the Hampden Company produced a line of heavy gold-plated watch boxes and adopted a plan of giving them away to dealers (they were worth about $15-20 per dozen at that time).

In the summer of 1878, Charles T. Higginbotham, the shop foreman, took out a patent for a new winding click. The *Sentinel* reported, "The American watches are fast taking the place of those in Switzerland, the superiority of the machine-made watch over a hand made being apparent." (7) At that time, the company employed "100 hands and turn out 40 to 50 watches daily."

FIGURE 7   *The Hampden Watch Company Factory, 1880*

In 1881, a new brick building was erected. It was 40 feet by 100 feet and had three stories. This brought the number of buildings used by the company to five, and by then it was producing 14 grades of movements--all full plate, except the 16 size "State Street." The number of employees grew to 400, and they could produce 400 watches per day. (8) A watch of the lower grades "passed through 650 different processes and one of a higher grade through 758 before being ready for the trade." (9) Yet, all parts were interchangeable. When the watches were completed, they were taken to the test room, where they were timed by electrical impulse from the observatory at Cambridge, Massachusetts. This process lasted three weeks to a month. This precision manufacturing enabled the company to offer a one-year guarantee against defects. Purchasers could receive a new watch or money back for a faulty timepiece.

FIGURE 8   *Hampden Movements*

Another Move

By 1883, the Hampden Watch Company was the leading industry of Springfield, Massachusetts, employing 600 persons--mostly skilled mechanics.   In that year, the company produced approximately 120,000 watches.   (Table 5) It was a surprise, therefore, when the company was sold in 1886 to the Dueber Watch Case and Manufacturing Company, from which it had been buying cases.   The person responsible for the merger, John C. Dueber, came to the United States when he was nine years old.   He was apprenticed to a watchmaker for five years in Cincinnati, Ohio, and by working nights making wedding bands, he accumulated enough liquid capital to start his own watch case business.   In 1863, he joined with Francis Doll to commence his case company in the Carlisle Building in Cincinnati. Dueber moved several times to Newport, Kentucky and back to Cincinnati, and in 1874, he finally moved to a new building at Washington and Jefferson Streets in Newport. (10) But for several

Table 5

HAMPDEN PRODUCTION

Year	Yearly Total	Grand Total
1877	60,000	
1878	80,000	140,000
1879	90,000	250,000
1880	110,000	360,000
1881	10,000	370,000
1882	75,000	450,000
1883	120,000	570,000
1884	110,000	680,000
1885	170,000	850,000
1886	100,000	950,000
1887	50,000	1,000,000
1888*	150,000	1,150,000

* Some of the dates after 1888 carry a "Canton, OH" name on the plates

Source: Cooksey Shugart, <u>The Complete Guide to American Pocket Watches</u> (Cleveland: Overstreet, 1981), p. 165.

years, he maintained a post office address in Cincinnati. The new location had 12,000 feet of floor space and 60 employees.

Because the company could not satisfy demand in that one location, Dueber built a new factory at Washington and Madison Streets, where silver cases were produced. The old building was then used exclusively to make gold cases. As business continued to grow, Dueber outgrew his two plants. Finally, in 1886, he incorporated the Dueber Watch Case and Manufacturing Company. The incorporators included Dueber and Joseph Daller, John Daller, and W. A. Moore. There were 200,000 shares of $10 each (total capital of $2,000,000).

To return to the main story, the Dueber company had been supplying cases to the Hampden company, but its biggest customers were Elgin, Waltham and Illinois. As the watch and case businesses became more competitive, "the big three" chose to form a watch case trust to regulate production. By this move, Dueber was boxed in. He did not wish to join the trust, and he could not expand any further in Newport, Kentucky. He chose instead to purchase the Hampden Watch Company of Springfield, Massachusetts. It was a good business venture. Hampden had $150,000 in cash and was paying a 10 percent dividend.

Dueber had intended to locate in Springfield, but was unable to find suitable land there. Since he could not expand at Newport, Kentucky, he advertised for a location to build new factories. He settled on Canton, Ohio, which donated $100,000 and 20 acres. Ground was broken on October 14, 1886 for the Dueber-Hampden factories. They were destroyed before completion, therefore operations were postponed until August of 1888. In the first year at Canton, the new Company turned out 600 watches, using 1,000 employees. Many of these came from the old factories, and it may be correct to say that when these persons left Springfield that it marked the official end of the Hampden Watch Company of Springfield, Massachusetts.

References

1. Charles S. Crossman, *The Complete History of Watchmaking in America*, Jewellers Circular and Horological Review, 1885-87, p. 108.

2. James W. Gibbs, *The Dueber-Hampden Story* (Philadelphia, 1954), p.6.   See also Percy Livingston Small, "Don J. Mozart," NAWCC *Bulletin*, VII (April 1957), 466-471.

3. Crossman, *Complete History*, p. 110.   See also W. Barclay Stephens, "The New York Watch Company," NAWCC *Bulletin*, IV (Feb. 1951), 279-287.

4. These names are important because they are part of the story of the founding of the Hamilton Watch Company.   Once again, it is difficult to reconcile lists of directors of watch companies. Gibbs includes James Abbe, James D. Brewer (Bauer?) and N. F. Leonard--names not found on the Crossman list.

5. Gibbs, *The Dueber-Hampden Story*, p. 7.

6. *Fitchburg Daily Sentinel*, January 30, 1878, p. 2.

7. *Fitchburg Daily Sentinel*, June 17, 1878, p. 2.

8. Henry G. Abbott, *The Watch Factories of America* (Reprint Adams Brown), p. 97.

9. Gordon A. Livingstone, "The Hampden Watch Company," NAWCC *Bulletin*, XIV (Feb. 1970), p. 156.

10. Gibbs, *The Dueber-Hampden Story*, p. 3.

CHAPTER 6:
E. HOWARD & CO.

At this point, we need to review some of the history that was related in Chapter 2. We saw there that Edward Howard used his mechanical genius in several other industries before trying the watch business. While he worked on butter and eggs, fire engines, and precision scales, he thought of manufacturing watches on automatic machines.

The Predecessor Companies

There is some question as to the precise year that Howard began making clocks or watches. He states, "I went in business myself as a clockmaker in 1840," but other sources give 1842 as the founding date of the E. Howard Clock Company. (1) Whichever date is correct, Howard made clocks for churches, offices and halls--both windup and electric. His shop at 15 Hawley Street, Boston, Massachusetts was only about 30 feet square. It was a meager beginning. But he soon moved and he moved often. The 1842 address tells us a surprising fact: that both Howard and Aaron Dennison were located at 69 Washington Street, without ever meeting.

In 1845, Howard moved to suburban Roxbury, where he built a new factory with Luther Stephenson and David P. Davis, who had been a fellow apprentice under Aaron Willard. They produced scales, precision balances and clocks. But Howard still thought of watches. To quote him again, watchmaking "on a comprehensive and systemic method was the result of many deliberations during the years 1848 and 1849, between Mr. Aaron L. Dennison and myself. . .Mr. Dennison being a watch repairer and myself a clock-maker, we made a good combination to systemize watchmaking, and to invent labor-saving machinery for producing perfect and interchangeable parts. With such views and intentions, we began the watch business in the spring of 1850, building a factory in Roxbury, Massachusetts." (2)

Table 6

BOSTON WATCH COMPANY CHRONOLOGY

Date	Item
1849	Howard and Dennison reach a partnership agreement to produce watches
1850 (Dec.)	Completed construction on new factory at East and Prescott Streets, Roxbury, MA Adopted Name of American Horologe Company
1851	Name changed to Warren Manufacturing Company
1852 (Nov.)	20-size, 8-day movements produced ("Howard, Davis and Dennison, Boston") production numbers about 1-17
1853 (Mar.)	18-size, 30-hour movements produced ("Warren, Boston"); production numbers about 18-120
1853 (Mar.)	18-size, 30-hour movements produced ("Samuel Curtis"); production numbers about 121-1000
1853 (Sept.)	Name changed to Boston Watch Company
1854 (Oct.)	Boston Watch Company moved to Waltham, MA.; produced 6 watches per day ("C.T. Parker," "Dennison, Howard and Davis" and "P.S. Bartlett, Waltham") production numbers 1001-5000
1856	Boston Watch Company produced about 100 movements ("Fellows and Schell")
1857 (Apr.)	Boston Watch Company sold at auction

For these ideas they were ridiculed. The notion of the superiority of hand work died slowly. Their friends considered their ideas insane and "told us we were crazy to attempt such an undertaking." But as Howard said, they were Yankees and they had the necessary "grit." The opposition to their plans meant "the financial problem was a hard matter to solve." Financial aid came from fellow Bostonians Samuel Curtis and Charles Rice.

With this help, they began a business which incorporated "a dozen distinct trades." under one roof--something that had never been tried before. The technical problems were enormous. Again, quoting Howard: "When it is understood that if any of the parts of a watch are one five thousandth of an inch thicker or thinner, longer or shorter, larger or smaller, than the proper sizes, the watch will not run well, it will be seen at once that the tools must be as near perfection as possible, to produce the exact and uniform sizes needed." (3)

If we can believe Harrold, he states flatly, "The founder of the American watch industry was Aaron Lufkin Dennison." (4) He too was apprenticed to a clock maker, but left Maine to travel to Boston to learn about watchmaking, which he did under Jubal Howe--who had been an apprentice of Luther Goddard in Shrewsbury, Massachusetts. In 1839, Dennison established a watch tool, supply and repair shop, and when it grew, he employed Nelson P. Stratton to assist him.

When Dennison and Howard started their collaboration, it was usually Dennison who tried to convince others to provide the financial means. One associate claimed that Dennison, "Could make more figures showing results than any man I ever met with." (5) But in fact, the primary source of initial capital came from Samuel Curtis-- Howard's father-in-law. Curtis was a dial painter, but he made his fortune producing mirrors. He never took an active part in the Howard-Dennison partnership, although he had enough faith in the business to invest $80,000 in it.

They began with a small room in the Howard and Davis factory (which made scales and balances, clocks and regulators). This space proved inadequate, so they purchased land in September 1850 at the corner of East and Prescott Streets for a new factory. The plant was completed in just over two months. It was 100 feet by 25 feet, with two stories. (6) The location was opposite the clock factory, which was a great convenience. Before the structure was completed, however, Dennison went to Europe to study watchmaking, English and Swiss styles, especially about making dials, gilding movements and constructing hair and main springs. He also sought materials that were not available in the United States.

The partners took the name of the American Horologe Company, and they made plans to produce their first watch--an 8-day with two mainspring barrels. They soon learned that this watch was not a reliable timekeeper, because the two barrels provided an uneven spring to the train. But Dennison insisted that the partnership pursue its goal of manufacturing an 8-day watch. He sought help from David and Oliver Marsh, master machinists, who designed such a watch in the fall of 1852. (7) D. S. Marsh made the model, but this plan was not satisfactory, hence they settled on 30-hour timepieces. But the bigger problem was how to produce these watches. The first plan was to have a master craftsman produce a single watch and to disassemble it and to reproduce the separate parts by machine. In practice, the pieces did not fit together. The state of technology in 1850 did not warrant this type of production. The machine did not exist that would produce identical parts; it had yet to be invented.

When production commenced, watch number 1 of the series was given to Howard and number 2 to Dennison. The watches were engraved "The Warren Mfg. Co." on the plates. Only 17 of these watches were made, and they were presented to officials of the company. (8) Watches that were numbered 18 to 1,000 had at least 3 names: Warren Manufacturing Company, Samuel Curtis and Fellows & Schell.

Howard and Dennison, who had hired Nelson Stratton to assist them in the technical aspects of watchmaking, found him to be both inventive and skilled, and he had worked earlier with the Pitkin brothers. Stratton believed that a watch with a single barrel "would never keep time over an eight day period," and he and Howard convinced Dennison to switch to a 30-hour movement.

Hauptman quotes Crossman concerning the necessary alterations, "The changes were to cut the barrel bridge in the center and put on a barrel bridge of the usual form, and, of course, throw aside the extra set of wheel and pinion, which had been used to make it run eight days. . .the third wheel, which previous to this had run under the center wheel, after the English style, was now raised to run over the center wheel." (9) Dennison continued to experiment in building machinery, but as Howard said later, "Mr. Dennison was a very fine watchmaker, but as a builder of watch machinery he was certainly not a success." (10)

In the fall of 1852, the company had produced enough machinery to commence production on the 30-hour movements. But when they developed gilding problems, Dennison sent Stratton to England to resolve them. On his return, about 100 30-hour momements were marketed in 1853 using the "WARREN, Boston" label. The company name had been changed to the Warren Manufacturing Company, after General Joseph Warren, who was born near the Roxbury plant and who was a hero of the Battle of Bunker Hill. (11)

In the spring of 1853, the "Warren Manufacturing Company," name was discarded in favor of the Waltham Improvement Company

FIGURE 9   *An 18 size, 15 Jewel Warren.   Sketch by*
George E.Townsend

and then the Boston Watch Company.   The next run of watches was
engraved "Samuel Curtis," and these were similar to English watches
with the fusees removed.   The watches were full plate and were called
18 size, after the new system Dennison had devised for this purpose.
He set one inch as 0 and divided the second inch by 30--making the
18 size 1 and 18/30ths inches in diameter.   In these movements, the
train was speeded up to a 17,200 beat with a larger mainspring. (12)

These 18-size movements of the Boston Watch Company were
full plate and can be recognized as the later model 57 of the Waltham
Watch Company.   These movements were numbered 1,001 to 5,000
and were labelled Dennison, Howard & Davis, C. T. Parker, and
P. S. Bartlett.   They had English cases and were marked on the
crown.   Although they looked British, they had some distinguished
features: a going barrel, case screws, and a quick train (later raised
to 18,000).   Some easily recognized names worked on these
movements.   For example, James L. Baker was in charge of

FIGURE 10 *The Boston Watch Company Factory, Roxbury, Massachusetts*

screwmaking machines (no small task, because each screw was cut, threaded. polished and blued automatically).

The Move to Waltham

The Roxbury location was not satisfactory. The land was rocky and high priced, and the employees could not afford to buy or build homes nearby. Dennison and Howard looked for an alternative. The best option seemed to be in the vicinity of Waltham. They chose the Bemis farm, which included 106 acres, but the land had recently been purchased by the publishing company of Little, Brown of Boston. James Brown, the publisher, agreed to sell the land in December 1853. William H. Keith handled the financial affairs. according to instructions from Dennison. They named the new firm the Waltham Improvement Company, with a common stock of $100,000 divided into $1,000 shares each. The Boston Watch Company purchased $30,000. The Improvement Company, as a land company. was given the contract for erecting the factory. The concrete building faced the river and measured 100 feet square. It was completed in October 1854.

The *Waltham Sentinel* states that this plant on the southern bank of the Charles River, one half mile from the Waltham train station "is the only watch manufactory on this side of the Atlantic; indeed may we say it is the only manufactory of its kind in the world." (13) The *Sentinel* went on to state that the Company produced "ten or a dozen elegant and excellent watches," daily, in silver cases ($30-50 each) and in gold cases (about double the price). They employed about 75 persons, mostly young men and young women and used a 12-horsepower engine to drive the equipment. More important, the *Sentinel* tells us that, "The whole force of the establishment is turned upon the manufacture of thirty-hour watches in hunter cases."

At that time, the old financial problems arose again. Curtis had invested $20,000 earlier, as had Howard, and now funds were short again. This time capital came from New York City. Fellows and Schell, of Maiden Lane, pledged an additional 20,000.

All business was done in the name of the Boston Watch Company, but the problem was not enough business. The company was averaging only 6 to 10 watches per day--not enough to remain solvent--and the panic of 1857 brought the company into bankruptcy.

Were They Howards or Walthams?

There has been a "friendly controversy' in the literature concerning the first and early watches produced at Roxbury and Waltham. Were they Howards or Walthams? In 1961, Hauptman surveyed the issue in an NAWCC *Bulletin* article, and he stated categorically that "these watches. . .are Walthams." (14) Others are not sure. The most convincing argument for calling them Howards came from George V. White. As he stated, "The first watch put out by the present company's predecessors in Waltham in 1854 was an 18 size '57 model' engraved Dennison, Howard & Davis and bore the serial number 1,000; and that all watches of this type bearing lower serial numbers are Roxbury made in Edward Howard's factory before watchmaking in Waltham was hardly considered." (15) However, although Howard could have claimed these watches as his own, he chose not to, because they were "far from his liking." White concludes, "From this it is plain that the Waltham company has no right to claim serial numbers of less than 1,000." There is further evidence that the early Dennison, Howard & Davis watches are really Howards. Watch number 1, which is in the Smithsonian Institution, is Edward Howard's personal watch; it is an 8-day with two mainspring barrels.

Beginning at the Bottom

In March of 1857, when the company was bankrupt, it assigned several watches to Charles E. Rice, who was a manufacturer of straw goods in Boston. The next month the firm was sold at a sheriff's auction to Royal E. Robbins and Tracy & Baker of Philadelphia for $56,000. Howard bid for his company, but the price went beyond his means. Some of the tools and materials were taken by Howard and he returned to the first factory. These included several unfinished movements, which were sold later as Howard & Rice models. As he said, he intended "to begin at the bottom and make all tools anew." (16) He was joined by 15 men, most of whom had worked for him in Waltham. The new company was the Howard Watch and Clock Company, with an authorized capital of $150,000, of which $120,000 was raised for production.

The above quotation not withstanding, Howard wasn't really at the bottom in 1857. Under the original mortgage given to the Boston Watch Company, Charles Rice had the right to certain watch materials and machinery (the exact amount is in dispute), but he took these "movable goods" to the old factory and allowed Howard to manage the

Table 7

E. HOWARD AND COMPANY CHRONOLOGY

Date	Item
1857	Howard and Dennison Company fails; Howard returns to Roxbury MA
1857	Howard Watch and Clock Company founded
1857 (Nov. 24)	George P. Reed patents stationary mainspring barrel
1858 (Dec. 11)	E. Howard and A. Howard purchase company from Charles Rice; Company renamed "E. Howard & Company"
1858	Production of 18-size, 7-pillar watches begins
1861 (Mar. 24)	The Howard Clock and Watch Company is incorporated
1863 (May 6)	Factory closed
1863 (Oct. 2)	Another new company started with Edward Howard as President
1865	George P. Reed resigns as Superintendent
1870	J.J. Pierson, of 15 Maiden Lane, New York, is employed as sales agent for company
1870	Howard develops new system for sizing watches: size A equals one inch, additional letters are 1/16 inch larger
1871	Production of Reed barrel is discontinued; new movement is marketed
1873	New plant completed
1881	The E. Howard Watch & Clock Company organized
1882	Edward Howard sells his personal interest in company for $81,000. Howard retires
1903	E. Howard Watch and Clock Company becomes E. Howard Clock Company. Howard name sold to Keystone Watch Case Company, of Philadelphia PA
1927	The "Howard" name is sold to the Hamilton Watch Company, of Lancaster, PA
1933	The Howard Clock Products Company is incorporated

plant for his (Rice's) benefit until December 11, 1858, when Howard purchased the property in association with his cousin Albert Howard, who also had been a clock apprentice. They settled all accounts with the creditors and began their new company, "E. Howard & Company." (17)

In the old Boston Watch Company, Howard had been Treasurer, and he spent most of his time filling financial gaps. It left him little time to think about watch design. The 8-day and 30-hour watches were promoted by Stratton and Marsh. They were essentially English, with modifications. Howard wanted to produce a superior American watch. As Small and Hackett stated, "Dennison had considerable skill and experience as a watch repairer, and watch and jewelry dealer in both retail and wholesale fields, but had none as a designer or maker of watches or watch machinery, nor as a factory executive." (18)

Reenter George P. Reed

In the new company, Howard took charge of the mechanical department, and he chose George P. Reed, who had been with him at the Boston Watch Company, to be his new assistant. Reed was always dissatisfied with the 30-hour movement, so he and Howard designed a totally new watch. They were assisted by N. B. (Napoleon Bonaparte) Sherwood. To produce this watch, there was need to design all new machinery and tools.

Reed is best known as the inventor of Reed's Patent Barrel, but his contribution to watchmaking is much larger. He was born in Grafton, New Hampshire in 1828, and at 18, he was apprenticed to a harness maker. While working on harnesses, he acquired a broken verge watch, which he attempted to repair by filing a part from a harness needle. He got the watch to run, and soon after he left the employment of the harness maker to work for Jacob Carter, a watchmaker from Concord, New Hampshire. Two years later (about

1849), he left for Boston, Massachusetts.

In 1854, Reed began working at the Dennison, Howard & Davis plant, in Roxbury, where he was put in charge of the pinion finishing room. It was there that he patented his new mainspring barrel on February 18, 1857. Because of the importance of this new device for reducing the damage from broken mainsprings, he was awarded a gold medal by the Massachusetts Charitable Mechanics Association.

When Edward Howard left Waltham in 1857 to return to Roxbury, Reed went with him. He stayed with the Howard firm until 1865, when he started his own business. On April 7, 1868, he patented a simplified escapement for a chronometer, and he built a shop in Malden to produce these watches, which he felt would be easier to repair than other chronometers. Reed also patented the whiplash regulator and the going barrel, which was used on Howard watches. The chronometer had two unusual features: a steel spring for locking and unlocking and an escape wheel with short teeth. Reed also produced a high-grade lever watch. But between 1868 and 1871, he produced only 100 movements, and to increase his output, he rented space in the old Tremont Watch Company building in Melrose.

Reed produced very few movements in his lifetime (which is why some are now valued at over $15,000 each). Those he did manufacture were 18 size, three-quarter plate, nickel and well designed. He sold them in Boston. Several writers claim that his most successful watch was made in 1862--the Monitor. It was supposed to be the first rotary made in the United States. But George Townsend contested this claim. He stated that the "Monitors" are actually watches produced by "A. H. Potter, Boston," and they are Tourbillons. As he said, "Reed had many patents but I haven't seen one for a rotary." (19)

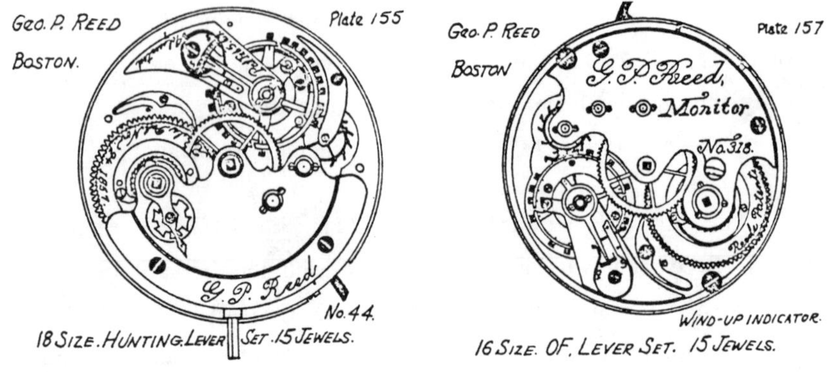

FIGURE 11   *The 18 Size G. P. Reed and the 16 Size Monitor. Sketch by George E. Townsend*

Within a year after starting, Howard was able to market his first new watches. They looked like the 18-size watches that were produced at Waltham, but they were entirely different. Because the top plate was made in sections, the watch required more pillars. Howard chose seven (four being customary). He used Reed's new stationary mainspring barrel. In a fusee watch, a malfunction of the mainspring caused the chain to break, but in a non- fusee watch, when the mainspring broke or released, it put a tremendous pressure on the wheels, usually causing damage to the train. Reed's new barrel was set in the pillar plate with the spring attached to the plate and the center wheel. If the spring broke, it simply unwound in the barrel without damage. Although Reed's barrel was called "Reed's Patent, November 24, 1857," it could not be patented, because it was an old idea. The patent covered the power retention device, not the barrel. In addition to the fame, Reed must have earned a good return on the invention, because Howard paid him one dollar for each movement containing the device. (20)

With his excellent team of designers, Howard brought out a number of features, some of which became permanent parts of watch manufacturing; others failed their tests. The fast train, "quick beat," was an immediate success. Raising the beat to 18,000 per hour made the watch more accurate. Other changes in design included the Mershon rack lever regulator, Cole's resilient escapement and the helical hairsprings.

Just as the new company was getting started, it was hit by the Secession Depression of 1860. It was difficult to sell fine, expensive watches. To survive, Howard sought more outside capital. On March 24, 1861, the Howard Clock and Watch Company was incorporated in Massachusetts, with a capital of $120,000. New Officers were elected: Albert Howard was chosen President, Edward Howard Treasurer and Superintendent and James W. Oliphant, Clerk. Other Directors included Henry W. Howard, W. S. Eaton, Solmon Piper and George Kingman.

But inventories kept piling up. The company borrowed $10,000 against their $17,000 value to pay wages and salaries. The management was reorganized again. George P. Tew replaced Albert Howard as President. But on May 6, 1863, the directors voted to close the factory and to dispose of the assets. This order, apparently, was not carried out; instead, on October 2, 1863, another new company was formed. This time Edward Howard became President, Albert Howard Treasurer and R. S. Lakin, Clerk.

David P. Davis, who had been associated with Howard since 1829, when they served as apprentices to Aaron Willard, Jr., left the company in 1865. It should be remembered that some of the earliest watches were labeled "Dennison, Howard & Davis." When Howard moved to Waltham, Davis stayed behind to continue working on scales, sewing machines and fire engines, but when Howard returned to Roxbury, their association resumed, although the extent of Davis's

commitment to watch manufacturing is not really known. He was
never an officer of a Howard company, and when he left, he became
a clock manufacture.

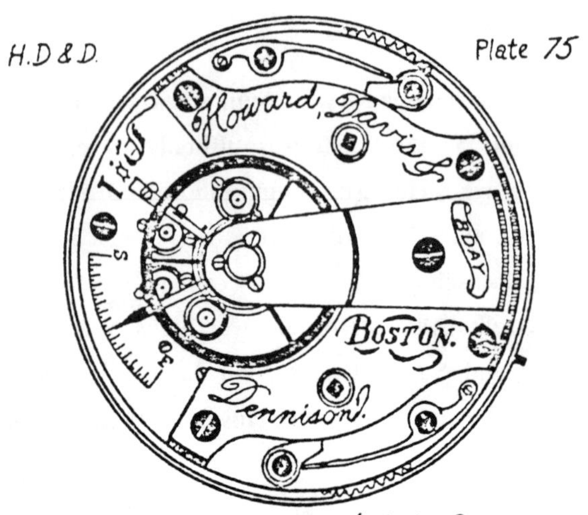

FIGURE 12   *The Two Barrel, 20 Size Howard, Davis &
Dennison.   Sketch by George E. Townsend*

In 1866, "American watches were introduced extensively into
London." (21) English watchmakers came to the United States to buy
American machinery.   Howard said of this, "It would sometimes seem
that Americans are altogether too good and accomodating, desiring to
let the whole world know what they do and just how to do it." (22)
But it was not all clear sailing for Howard or the other Americans.
Interchangeability of parts "was not found to be practical in the finer
parts; never have been to this day, and never will be." (23) Although
this pessimistic assessment was not borne out in time, it did describe
the technical state of the market in which Howard was trying to
penetrate.

Resolving Technical Problems

George Reed resigned as Foreman and Superintendent in 1865 to go to England. After him, there were several personnel changes in these posts. J. A. Dawson served, then F. A. Jones, who later tried to make watches for the International Watch Company of Chaux-de-Fonds in Switzerland using American machinery. Jones left Howard in 1868 and was replaced by J. H. Gerry, formerly Superintendent of the Springfield Watch Company. Gerry invented a stem-winding attachment that, with modifications, could be used with the Reed Patent barrel. In 1871, however, the Reed barrel was discontinued in favor of Gerry's, because it could be used in both stem and key winding watches. The "resilient" hairspring was another technical problem: to eliminate banking pins, Howard had the pallet jewels bank against a shoulder on the back side of the escape wheel. These were troublesome, most being returned. and they were replaced with the usual escapement.

The new movement (first marketed in 1871) used a micrometer regulator, which Reed had patented in 1870. but Charles Fassolt, of Albany, New York, who had previously patented a smaller regulator, claimed a patent infringement. The suit was settled with Howard paying Fassolt $1,000 for the use of the regulator. (24)

The Howard dials reflected the high standards of Josiah Moorhouse, foreman of the dial department at the Howard factory for many years. Moorhouse had been trained in London, where porcelain dial making was a high art. He emulated the master craftsmen of London, who often signed their names on special dials. His dials had five minute numerals (red and black), and he often added small flourishes to the Howard name. Some dials were signed and dated and showed the name of the persons who who ordered them. (25)

The accuracy of the watch depended, in part. on the quality of the jewels. The early American movements had large colorless beryl cap jewels and chrysoberyl (chrysolite) plate jewels. This European technology was brought to the United States by Dennison when he returned from his first trip there, during which he was seeking watch materials. He brought with him a jewel maker named Sibley, who experimented with several types of materials for making jewels.

We will return to jewel technology in the next section, but first we must treat another problem. As the public came to prefer stem-wind watches, Howard found itself with mostly hunting case movements. The solution was the Abbott Stem Wind Attachment. It was used on both Howard and Waltham watches until about the turn of the twentieth century. It came in two styles and it was fitted by jewelers.

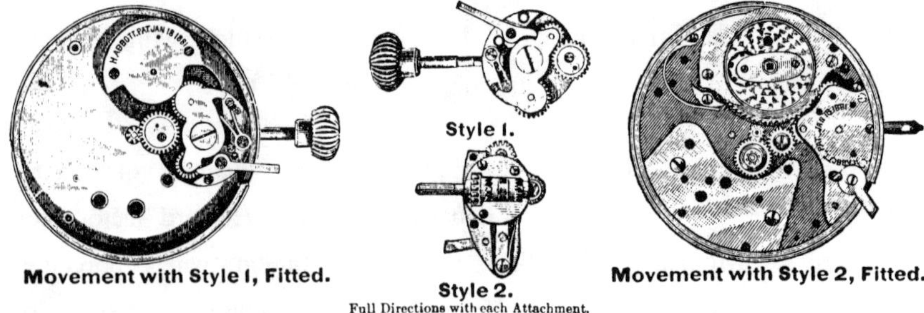

FIGURE 13 *Abbott's Stem Wind Attachments. Taken from the 1895 Otto Young & Co. Catalog*

Enter Napoleon Bonaparte Sherwood

It was N. B. Sherwood who revolutionized the making of watch jewels. He invented tools and machinery for making improved jewels, and he described the whole process in his book, *Watch and Chronometer Jeweling* (George H. Hazlitt, 1869).

Sherwood was born on November 11, 1822, but little is known of his early life. Small thinks Sherwood was born in Albany, New York and was the son of a carpenter. (26) At age 12, the boy entered Albany Academy, preparing for the study of medicine, but for unknown reasons, he terminated these studies. Sherwood and a friend journeyed to Chicago, where they arrived in the middle of the depression of 1837. After working briefly as a draftsman, he moved again, this time to Pennsylvania. He opened a drug store (using his knowledge of medicine), and curiously, he began his study of watchmaking. Shortly after, he married a Miss Mary Van Valkenberg, of Albany, and they traveled to Jefferson, Ohio, where he opened a watchmaker's shop. He supported himself and his wife by teaching a class in mathematics. But Jefferson could not support a watch shop, and it had no machine shop, so again they moved, back to New York.

Before leaving Jefferson, Sherwood met George B. Miller, who later founded the first watch magazine in the United States, the *American Horological Journal*. Miller must have had a large influence on Sherwood. But, first, he drifted again, this time working as a traveling salesman. Somehow, he met William H. Willson, of Howard & Davis, of Roxbury, Massachusetts, who was selling sewing machines, and who was the son-in-law of Samuel Curtis, who had just financed the establishment of the Boston Watch Company. Willson invited Sherwood to visit the factory, and so he met Edward Howard. Their discussions were quite compatible, and in 1854, the Sherwoods moved to Waltham. As a mechanically-inclined draftsman, Sherwood was left to his own imagination. Because his traveling friend in Chicago had been employed as a jeweler, Sherwood was assigned first to the jewel department.   Sherwood found the employees opening jewels by hand, so he first invented a lathe for this function; then he mechanized the entire process.

Sherwood broadened the range of jewels that functioned well in the early Howard watches.   Kleeb tested the stones in these movements, and he found ruby, sapphire and chrysolite jewels on the plates and ruby, sapphire and garnet stones on the pallets.   Later acquamarine and diamond were used.   He did not evaluate the relative merit of these stones, but Sherwood wrote, "In our estimation chrysolite is the most valuable of all stones for jeweling."   (27)

Enter Jonas G. Hall

We digress a bit to introduce a curious figure in United States horological history--Jonas G. Hall.   He has been rated among "that famous quartet of men who did so much to make machine watch-making possible--A. L. Dennison, J. G. Hall, E. C. Bingham, and N. P. Stratton."   (28) Yet, he is relatively unknown.

Hall was born on June 24, 1822 at Calais, Vermont.   He began as an apprentice in 1836 at Montpelier, but he left Vermont to work with Maurice O'Connell, at 226 Ann Street, Boston, Massachusetts. Hall worked on marine chronometers and pocket watches for ship captains for 8 years and then he returned to Montpelier, where he established a partnership with Ira S. Town.   But he left in three years to start his own business.   In the 10 years between 1848 and 1858, Hall produced 65 lever watches and a marine chronometer.

In 1859, Hall returned to Boston and worked with the American Watch Company under Aaron Dennison.   Hall designed the first ladies watch manufactured by that company (a 10 size P. S. Bartlett and A. T. & Co.)   But the peripatetic Mr. Hall kept moving.   In 1862 and 1863, he was in Montpelier; from 1864 to 1865, he was at the E. Howard & Company plant in Boston; from 1866 to 1868, he stopped to work for the Tremont Watch Company and finally, from 1869 to 1871, he returned to the E. Howard & Company. Hall did settle down, at Roxbury, Vermont, where he made a mark as a foremost maker of watch tools. He invented staking tools, which achieved a world-wide reputation. When he died, in October 1891, his manufacturing company was sold to F. L. Herrick, of Springfield, Massachusetts.

Marketing Watches

Originally, the Howard watches were sold by Fellows & Schell, of 23 Maiden Lane, New York. When the second company was formed in 1863, a New York syndicate of five firms took over the marketing function. They invested $5,000 each, for which they gained control of the entire production of the firm. This system prevailed until 1870, when Howard decided to have his own agent and store in New York City. J. J. Pierson, of 15 Maiden Lane, assumed this responsibility. He soon moved to 552 Broadway and in 1886 to 41 Maiden Lane. The Boston store at 378 Washington Street was managed by Albert Howard.

In 1870, Howard further differentiated his product. Taking a cue from his old partner, Howard developed a unique system for sizing watches. He started with the letter A for one inch in diameter and added a letter for each 1/16th inch. He started making N size watches (1 3/4"), which were slightly larger than the usual 18 size watch. The L size (1 11/16"), which was almost equal to a 16 size, was introduced in 1871 as the first nickel movement, and a ladies watch (size C) appeared in the following year.

It should be noted that all E. Howard movements left the factory uncased. Some distributors cased the watches before sale; others sold just movements. But many of the early Howards (up to approximately 6,700) had cases of the Charles E. Hall & Co., and these are considered by many persons to be "original." These cases were marked "CEH & Co." or "EH & Co." Other manufacturers, who made cases for Howard, include: Crosby & Mathewson (C & M), J. M. Harper (J M H), D. T. Warren (D T W & Co.), Warren & Spadone (W & S), Fellows & Co. (F & Co.) and Wheller Parson & Co. (W P & Co.). (29)

With these improvements, the company seemed to prosper. At the annual meeting in 1869, the treasurer was authorized to buy land for a new factory. The plant was completed in 1873 by Henry W. Howard, a contractor who was the eldest son of Waters Howard. This was not a propitious time to expand; the depression of that year was the most severe of the nineteenth century. Jay Cooke and Company failed and many other financial houses followed. The New York Stock Exchange closed down from September 18 to 30. Hard times came to manufacturers and merchants.

The company survived, but it was buffeted continuously. Thomas Howard died in 1877, and in the spring of 1879, an associate firm, with which Howard was linked closely, also failed. This limited Howard's operation until December 1, 1881, when the company was reorganized again. The new company was called the E. Howard Watch & Clock Company, with an increased capital of $250,000. The new investors purchased the plant and all the assets of the old company. Edward Howard was elected President; Charles J. Hayden, Treasurer and Business Manager and Albert Howard, General Superintendent. In early 1882, the company purchased Edward Howard's personal interest in the business, and he retired (he being in his seventies at the time). Howard's share was $81,000. The new managers continued to make watches marked "E. Howard & Co." They were led by Samuel Little, who replaced Edward Howard, and the other Howards. However, the Howards were diminished by one when Albert Howard died. He had spent his entire lifetime in the clock and watch business.

In reality, the Howard Company made few changes in its watches between 1870 and the end of the century. The watches were high grade, but very unconventional. Many were unsuited for the railroad time service. As a consequence, by 1900, sales were at a standstill, and things remained as described above until 1903, when the Howard name was transferred to the Keystone Watch and Clock Company, of Philadelphia. The E. Howard Watch and Clock Company became the E. Howard Clock Company. It completed all its work in progress and converted to making clocks. Only the name "Howard" could be put on the dial and "E. Howard Watch Company" on the movement.

The Keystone Watch Case Company was the product of Theophilus Zerbrugg, of Riverside, New Jersey. He first combined a number of small watch case companies near Philadelphia to form the Philadelphia Watch Case Company. This company became Keystone. In February 1903, Zerbrugg incorporated the E. Howard Watch Company in New Jersey. He was its President. E. R. Snow, Secretary and C. M. Fogg, Treasurer. (30)

In March 1903, the Keystone Company purchased the land and buildings of the U. S. Watch Company at Waltham, Massachusetts. By 1905, it was producing an entirely new line of "Howard" watches under the name of E. Howard Watch Works. These were 17-23 jewels, and they were fine timepieces, but they were quite different from the earlier "E. Howard & Co." watches.

Are They Howards or Walthams (Again)?

Before the new E. Howard Watch Company produced its own watches, it finished and sold watches made for it by the Waltham Watch Company. They were 16 size, 23 jewels, and known as the 1899 model. They had gold trains and several Waltham features (triangular hairspring stud, pendant setting mechanism).

While the Howard company was being sold (1902), several Howards were made at Waltham. Borg illustrates one of these, which was Howard number 1,005,117 or Waltham number (12)005,117 (the 12 was omitted). Ehrhardt describes this watch as a bridge model, 23 jewels, Special grade and position. Jefferson pictures another of these Howard/Waltham hybrids, Howard number 1,010,401 and Waltham number 12,610,401--a 19 jewel, 1899 model, with hunting case, also Special grade. (31)

The New Howards

The first of the all-new, post-1903 Howards was the 16 size 1905 model. It was unlike earlier Howards, but similar to other American watches of that type (the Waltham 1888 model and the 16 size Hamilton). It was offered in three-quarter plate, open face and hunting type, with nickel finish and a Breguet hairspring. The first runs were 17 jewels, but additional ones were made in 19 and 23 jewels (bridge movements).

Figure 14 illustrates two 16 size, 17 jewel Howards. On the left is number 913,624 (1907). It has an I W C Co. nickel case (4,757,870. The watch on the right is a bridge movement, adjusted, "Pat'd '08," number 1,092,518 (1919), and it has a 14K K W Co. case (117,858).

These new 16 size Howards were designed by Joseph A. Freund, an Austrian immigrant, who was born in 1852. As Borg states, "Little is known of this man." (32) He served an apprenticeship in his youth and emigrated to New York City, where he became a United States citizen. Sometime between 1907 and 1910, he started his association with the E. Howard Watch Company as a designer. He served as a general manager and some-time inventor--one of the most useful of which was for a free-sprung

FIGURE 14   *Two New Howards*

balance.   This balance was used on "Edward Howard" watches after 1912.   Freund also designed watches for the New York Standard Watch Company, and naturally, the two companies produced watches with some similarities.   This was particularly true for the Howard 10 and 12 size bridge models and the New York Standard 12 size.   In fact, in some instances, there was a confusion between the models, especially since some parts were interchangeable.   But, as always, workmanship at the Howard plant was of a much higher quality.

The next in the line of new watches was the 12 by 14 "thin model," which had 14 size plates and 12 size bridges.   These have always been called 12 size.   They were pendant set, with safety barrels, and had 17, 19, and 21 jewels.   This watch was notable because it incorporated another Freund invention--the positive pendant set mechanism, that became standard after 1911.   Although the Swiss had used a positive setting mechanism since the 1850's, the Americans used the Duane negative set pendant after 1882.

Howard next introduced its new 16 size "Edward Howard," as a high quality watch for the "wealthy gentleman." The patent for this watch was granted to William H. Ebelhare and Joseph A. Freund on June 4, 1912. Only 250-300 of these movements were produced, which were sold for $350. This was Howard's answer to Waltham's Premier Maximus. But Borg maintains that the Howard was designed for a completely different reason. The Maximus was a conventional Waltham with a prestige finish; the Howard was a completely redesigned watch, with new bridges, plates, and screws. And the watch was hand finished in all parts. It had an 18 carat case, blue sapphire jewels, and came in an ebony presentation case. Because of its fine workmanship, it was given a new numbering system, starting at 1.

The last of the new Howards was the 10 size thin model, that continued the trend towards smaller and lighter watches. Borg compares this redesigned model in importance to the 18 size 1857 Waltham, because the 10 size Howard was a revolutionary watch, designed by Freund. (33) It had no pillars. Instead the movement used a thin train bridge and a thick pillar plate. This concept was not new, however, because it was incorporated as a feature in Charles Vander Woerd's patent number 95,547, of October 5, 1869. By drilling them together, several operations were eliminated, and, more important, the stem holes were guaranteed to line up. The watch was relatively simple: it was open face, had a common (not a micrometric) regulator, nickel silver finish and pendant set, and it came in 17, 19 and 21 jewels. The watch was also much easier to repair.

The revolutionary aspect of the Howard 10 size watch was really adopted in making wrist watches, especially the Hamilton and Elgin /0 sizes. Not only was the 10 size Howard easier to construct and to repair, but as Borg stresses, the 21 jewel model was the only American watch of its time to qualify for the Geneva Stamp, although

American watches could receive that stamp of approval.

The old "E. Howard & Co." continued to make escapements for clocks and safe works. In 1930, Keystone sold some of the land and buildings at Waltham to the E. Howard Clock Company, which was incorporated in 1927 to take over the business of Howard & Davis. The "Howard" name was sold to the Hamilton Watch Company of Lancaster, Pennsylvania, but the Howard Clock Company was given the contract to complete watches then in production. It also made parts for Howard watches, as well as certain instruments.

Raymond S. Wilder, who was the manager of the Howard Clock Company, resigned in 1933 to start his own business. In the next year, the Howard Clock Products Company was incorporated, with Harold C. Keeman as President and William J. Elton, Treasurer. they reached an agreement with E. Howard Clock Company to lease its plant with an option to buy. The two officers involved in this transaction, Keeman and Elton, had both worked for the Howard Clock Company. Elton spent his entire work career at Howard, but Keeman began as an apprentice with the Waltham Clock Company. He was the superintendent of that company until 1924, when he joined the E. Howard Watch Works, where he became secretary and general manager, until he left in 1928 to work in the watch and clock repair business for his own account. The Howard Clock Products Company developed a successful business of making small parts and gear trains and complete clocks, after the old Howard design.

Some Interesting Howard Watches

While all Howard watches are considered interesting, some are more so. One contained the ingenious Mershon Patent regulator, which was devised by an Ohio jeweler to fit an imported English watch. Watch number 2940 (illustrated by Cuss in Plate 150) shows this devise. It is a rack, which is fitted to a regulator arm that is pivoted around the central hand- setting lever. Moving the arm along the index provides a very slight movement of the rack, thus enabling the regulator to adjust the watch closely. (34)

George Christiansen relates his experience in cleaning a "rare Howard watch." It was a 15 jewel movement, with serial number 6518. Dana Blackwell explained that this N size Howard was one of several unique Howards. In this case, the Reed Patented Barrel was, in fact, no barrel at all. The main spring was attached to the pillar plates. This design was Reed's answer to broken main springs that often damaged the wheels. This particular watch was completed on March 23, 1865 and was adjusted by Reed. (35)

As another entry in the unusual Howard category, we offer Howard watch number 12,772, which passed final inspection at the factory on May 17, 1867. This was probably a Series III, size N, with Reed's Patent and a helical hairspring. What made this watch special was that it it was carried for 42 years on 18 vessels of the United States Navy by Admiral Charles D. Sigsbee. The watch went down in Havana Harbor on February 15, 1898, when the United States Maine was sunk. It was recovered by a navy diver and presumably restored to its former elegance. What is astounding is that this episode is the *third* time that the watch was under water, although in the first two occasions, it was submerged for only minutes. (36)

We cannot leave this section without mentioning the "Edward Howard." When Waltham introduced its Premier Maximus, it was considered to be the finest timepiece in United States history. The Howard Company had to meet this competition. The "Edward Howard" was its answer. First made in 1912, it was embellished with an 18 carat case and a presentation box. The 23 jewel movement had blue-sapphire jewels, free-sprung balance, chronometer adjustment and frosted bridge plates. Howard claimed that it was the "finest watch ever produced in this country or any other country."

FIGURE 15   *The Edward Howard, Number 250 (1920)*

Howard Production Figures

It is very difficult to ascertain the exact production figures for E. Howard & Co., because in many cases the production runs began with very high numbers, and the numbers often were not produced sequentially.   Howard began producing seven pillar watches. According to Shugart, he manufactured about 1,500 of these, then he shifted to three-quarter plate movements in 1860.   Most of these had the Reed Patent Barrel, of which about 36,000 were produced. Shugart's output statistics up to 1903 are given in Table 8.   But these are at odds with Townsend's data.   Shugart tells us that Howard produced 109,650 watches from 1858 to 1903 (page 17), but Townsend states that Howard "made approximately 900,000 watches from 1857 to 1903. A reader of the NAWCC *Bulletin* puts us deeper into this statistical thicket when he writes, "I have read that E. Howard & Co. did not make over 160,000 watches under his name."   (37)

Table 8

HOWARD PRODUCTION

Date	Serial Numbers	Size	Series	Production for Period	Cumulative Total
1858	1 - 1,980	18	I	1,900	1,900
1860-1861	2,001 - 3,000	18-N	II	1,000	2,900
1861	3,001 - 3,400	18-N	K	100	3,000
1860-1870	3,500 - 27,500	14K	III	25,000	28,000
1870-1880	30,000 - 45,000	N-18	IV	10,000	38,000
1870-1900	50,000 - 70,000	L-16	V	20,300	58,300
1870-1900	100,000 - 105,200	G-6	VI	5,200	63,500
1880-1900	200,000 - 226,500	N-18	VII	26,000	89,500
1895[*]	228,000 - 230,000	N-18	VII	2,000	91,500
1885-1900	300,000 - 308,600	N-18	VIII	8,600	100,100
1895[*]	309,000 - 310,000	N-18	VIII	900	101,000
1890-1895	400,000 - 405,000	N-18	IX	5,000	106,000
1890-1900	500,000 - 501,300	I-10 I-12	X	1,300	107,300
1895[*]	600,000 - 601,100	I-10 L-16	XI	1,100	108,400
1895[*]	700,000 - 701,300	L-16	XII	1,250	109,650

[*]Three-quarter split plates with 17 jewels

Source: Cooksey Shugart, <u>The Complete Guide to American Pocket Watches</u> (Cleveland, TN: Overstreet, 1981), p. 171; George E. Townsend, <u>Almost Everything You Wanted to Know about American Watches and Didn't Know Who to Ask</u> (Vienna, VA, 1970), pp. 20, 25.

Figure 16 shows the four most popular Howard watches, according to Townsend's reckoning. They include the Series VII, N size, 26,000 produced; the Series III, K size, 25,000; the Series V, L size, 20,300; and the Series VIII, N size, 8,600.

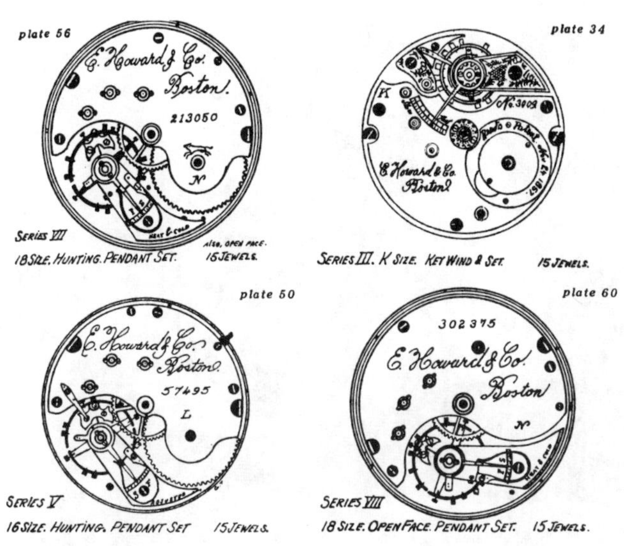

FIGURE 16   *The Four Most Popular Howards.   Sketches by George E. Townsend*

When Keystone took over E. Howard & Co. in 1903, there was more confusion over series and serial numbers. The Series III and IX is an obvious example. Their plates were identical, but "they were different in the damaskeening of their top plates." Series IX had a checkerboard design; Series III a circular design. To add to the confusion, there is another 17 jewel, three-quarter plate, 16 size Keystone Howard--the 1905 model. It, too, is identical to the Series III and IX models, but the damaskeening has slanted bands. (38)

Finally, we are left with the Abbotts. Are they Howards? These "Sure Time Watches" were made by the E. Howard Watch Company (Keystone). They were 16 size, 17 jewel, adjusted, three-quarter plate movements. They had gold or silver dials, ruby jewels and gold cups, and they were sold with nickel cases. The watch pictured in Figure 17 is number 975,590. It is fitted with a yellow gold filled Philadelphia case (9,079,098).

FIGURE 17   *An Abbott "Sure Time"*

References

1. See Edward Howard, "American Watches and Clocks," in Chauncey M. Depew, *One Hundred Years of American Commerce* (New York: D. O. Haynes, 1895), p. 541.

2. Howard, "American Watches and Clocks," p. 541.

3. Howard, p. 542.

4. Michael C. Harrold, *American Watchmaking: A Technical History of the American Watch Industry, 1850-1930* Supplement, NAWCC *Bulletin*, Spring 1984, p. 16.

5. E. Tracy to C. N. Thorpe, NAWCC *Bulletin*, April 1949, p. 603.

6. Wesley R. Hauptman, "The Boston Watch Company," NAWCC *Bulletin*, X (Oct. 1963), p. 924.

7. Chris Bailey, *Two Hundred Years of American Clocks and Watches* (Englewood Cliffs: Prentice Hall, 1975), p. 196.

8. Harrold, *American Watchmaking*, p. 18.

9. Hauptman, "The Boston Watch Company," p. 926.

10. Charles S. Crossman, *The Complete History of Watchmaking in America* (Adams Brown Reprint of Jewellers and Horological Review, 1885-87), p. 14.

11. Bailey states that Dennison was working on a 36-hour movement, but this must be a typographical error. (p. 196)

12. We should probably trust Harrold's account, because it is later and because Hauptman stated that he never saw a Warren-engraved watch.

13. *Waltham Sentinel*, March 13, 1856.

14. Wesley R. Hauptman, "The First Waltham. . .or Are They Howards?" NAWCC *Bulletin*, X (Dec. 1961), p. 27.

15. Quoted in Hauptman, "The First Waltham," p. 27.

16. Howard, "American Watches and Clocks," p. 542.

17. See Charles S. Crossman, *The Complete History of Watchmaking in America*, p. 55.

18. P. L. Small and F. E. Hackett, "E. Howard: The Man and the Company," NAWCC *Bulletin*, Supplement, 1962, p. 11.

19. George E. Townshend, "Answer Box," NAWCC *Bulletin*, XX (Aug. 1981), p. 410.

20. Crossman, [The Complete History,] p. 56.

21. George Lewis Dyer, *The Story of Edward Howard and the First American Watch* (Boston: E. Howard, 1910), p. 19.

22. Howard, "American Watches and Clocks," p. 542.

23. Small and Hackett, "E. Howard." p. 13.

24. Crossman, *The Complete History*, p. 61.

25. D. J. Blackwell, "Answer Box," NAWCC *Bulletin*; XV (Feb. 1973), p. 999.

26. Percy Livingston Small, "Napoleon Bonaparte Sherwood," NAWCC *Bulletin*, VI (Dec. 1953), 24-28.

27. Alvin A. Kleeb, Watch Jewels of the Past," NAWCC *Bulletin*, X (Apr. 1962), 191-198.

28. From the obituary of N. P. Stratton, quoted in Albert O. Dodge and George G. Lucchina, "Jonas G. Hall, 1882-1891," NAWCC *Bulletin*, XVIII (Oct. 1976). p. 437.

29. George E. Townsend, *E. Howard & Co.* (Kansas City: Heart of America Press, 1982), p. 4.

30. Edmund L. Sanderson, *Waltham Industries* (Waltham: Waltham Historical Society, 1957), p. 107.

31. See Arthur N. Borg, "The Howard Ten Size Watch," NAWCC *Bulletin*, XII (Aug. 1967), p. 946 and George J. Jefferson, "Another Howard (Waltham Product) Watch." NAWCC *Bulletin*, XIII (Oct. 1968), p. 548.

32. Borg, "The Howard Ten Size Watch," p. 948.

33. Borg, p. 954.

34. Theodore C. Cuss, *Cammerer Cuss Book of Antique Watches* (Suffolk: Baron Publishing, 1976), p. 235.

35. Dana J. Blackwell, "Answer Box," NAWCC *Bulletin*, XXVII (Oct. 1985), pp. 605-606.

36. See the NAWCC *Bulletin*, XIV (Aug. 1970), p. 437.

37. F. A. R. Gonzales, "Answer Box," NAWCC *Bulletin*, XII (Feb.

1966), p. 158.

38. Robert A. Lavoie, "Keystone Howard Series 3 or 9?"   NAWCC *Bulletin*, XXVIII (Feb. 1986), p. 81

The Half Breed Watch

The Tremont Watch Company produced what Hauptman called a "Half-Breed" watch. This, too, involved the incredible Mr. Aaron L. Dennison, who, in his time, seemed always to be anywhere where a watch company was being formed. As we stated in Chapter 2, Dennison was born in Freeport, Maine, but he went to Boston to learn watchmaking. In the years from 1833 to 1848, he drifted much, moving to New York City, Springfield, Massachusetts and back to Boston.

On returning to Boston, Dennison met Edward Howard, and their meetings resulted in the formation of the Howard and Davis shop and the American Horologe Company. The technical and financial problems of this company and its successors have already been detailed in Chapter 6. The company went through several reorganizations and name changes until 1857, when the Boston Watch Company failed.

Howard returned to Roxbury to form his Howard Watch and Clock Company. According to him, "Mr. Dennison was employed by the new company (Waltham) for two or three years." (1) Dennison was marking time waiting for his next entrepreneurial opportunity. It came in 1864, when he visited his old friend A. O. Bigelow, who was with the watchmaking firm of Bigelow, Kennard and Company, of Boston. Bigelow had been an advisor to Dennison since they met in the 1840's. It was not difficult, therefore, for Dennison to convince Bigelow that there was a large potential demand for a good machine-made watch. Dennison emphasized that Waltham could not meet the rising demand engendered by the American Civil War. If he "could succeed in placing a good movement on the market at a reasonable price, there would be a ready sale." (2)

Table 9

TREMONT (MELROSE) CHRONOLOGY

Date	Item
1864	Aaron L. Dennison visits A.O. Bigelow, a Boston, MA watchmaker to discuss formation of new watch company
1864 (May)	Tremont Watch Company incorporated with $100,000 capital--$50,000 paid in
1864	Watch factories opened in Zurich, Switzerland and Boston, MA
1864	Dennison goes to Berne and Zurich, Switzerland
1865 (May)	First "Half breed", 18-size, key wind, fully-jewelled watches produced
1866	Company moves to Main and Tremont Streets, Melrose, MA
1866	Bigelow resigns as Treasurer
1866	Dennison resigns as Superintendent and Director of the Tremont Watch Company
1866	New "all-American" 18-size movement produced
1868	Company out of funds; Factory closes
1868	Dennison tries unsuccessfully to sell Swiss factory
1870	Dennison returns to Boston to revive company--effort fails
1870	Dennison returns to Birmingham, England; Sale of Swiss Factory completed to Anglo-American Watch Company
1870-1895	Dennison manufactures watch cases
1895 (Jan. 9)	Dennison dies in England

In the spring of 1864, they formed the Tremont Watch Company--a stock company with initial capital of $100,000 in 10 shares, of which $50,000 was paid in. Dennison was made Superintendent and Bigelow Treasurer. The company office was at Washington and Franklin Streets, and the machine shop was on Hanover Street in Boston, Massachusetts. Charles P. Crafts was foreman of the shop.

Because of the high price of labor in the United States and because of the lengthy process involved in starting a watch factory (which Dennison had already experienced before), he conceived of an uncommon plan to produce watches quickly. He believed he could produce his first movement in twelve months. The plan involved starting two plants at the same time--one in Boston and one in Switzerland. The Swiss plant would produce the trains, balances and escapements; the Bostonians would make the plates and barrels and other parts needed to assemble a complete movement.

In addition to Crafts, who was general foreman of the Hanover Street shop, the company employed:

D. B. Bingham	Assistant Superintendent
John Polsey	Foreman--Plate room
Charles Byam	Foreman--Flat steel room
Osmore Jenkins	Master watchmaker
Daniel F. Leary	Foreman--Jewelling department
Andrew Brush	Gilder and patternmaker

While D. B. Bingham ran the Boston plant, Dennison left for Switzerland with a man named Godfried, who acted as assistant and interpreter. For reasons unknown, they left Berne and went to Zurich, where they expected to set up their European headquarters to finish all watch materials.

Dennison projected output figures of 600 trains per month, but the Swiss thought his goal was impractical. In fact, the Swiss factory was able to supply parts at a much faster rate than projected and faster than its Boston counterpart.

As Dennison promised, the first movements were on the market in about a year. They were 18-size, key wind, fully-jeweled and were engraved "TREMONT WATCH COMPANY, BOSTON" on the plates. Crossman rated them very highly. As he wrote, "They gave good satisfaction and were probably as good as any on the market at that time for the price." (3) The company "was able to sell every movement they produced. . . about 400 per month." (4)

The watch shown in Figure 18 is a standard 18 size Tremont, number 9288, made in 1865. It has an original Dueber coin silver case (756,235)

FIGURE 18   *An 18 Size Tremont*

Now Melrose

The Civil War business gave the Tremont Watch Company a

sense of false prosperity, and acting on this information, the Board of Directors (with Dennison absent) voted to move to Melrose--a suburb of Boston--where two members of the Board had property. This move was designed as a complete change of plans. The company would cease making parts in Switzerland and instead make complete movements in Melrose. When the vote was approved, Bigelow resigned as Treasurer, and he was replaced by Edward S. Philbrick. an architect from Boston.

The new factory was a frame building on Main and Tremont Streets. It was 50 feet by 100 feet, two stories, and it was remodeled to serve the watchmaking functions. The removal to Melrose did not please Dennison. Since he was in Switzerland, he had to argue his case by mail. He knew from his Waltham experience how difficult it was to make machinery which could produce interchangeable parts. And he did not believe that the Americans could produce parts as cheaply as the Swiss. Dennison thought that this would force Tremont to raise prices, which would jeopardize its competitive position with Waltham.

Dennison's view did not prevail, so he withdrew in 1866 as director of the watch company, and instead did work for it under contract. He assisted in the production of a new 18-size movement, which was engraved "MELROSE WATCH COMPANY, BOSTON." The introduction in 1866 of these new movements ended the production of "half-breed" watches. The new model was cheaper and was seven jewels instead of fifteen. But Hauptman states that, "I have seen or heard of over a dozen Melrose movements and they all have eleven jewels." (5) Some movements have Melrose dials; the name was intended to indicate a less expensive model. The company continued its Tremont watch, which was a higher grade, full-jeweled and more expensive, but few were made, because true to Dennison's warning, it was not able to compete effectively with the Waltham

Watch Company and the newly- formed National Watch Company (Elgin).   Since the $50,000 of paid-in capital was used up, the directors issued a call for an additional $150,000, which was authorized initially by the corporate charter.   When the stock holders did not respond to this call, the factory was closed in 1868.

A Melrose Watch is shown in Figure 19.   It is 18 size, 15 jewels, serial number 31,487, with the name Tremont on the dial. The hunter case is yellow gold filled, G W L 970.

FIGURE 19   *A Melrose Watch*

Meanwhile, the directors asked Dennison to attempt to sell the plant in Switzerland.   He failed at this task, but he remained in Switzerland until 1870, when he returned to Boston. He made a feeble attempt to revive the Tremont Watch Company, and failing this, he tried to organize yet another company.   When this too failed, he returned to England, this time to Birmingham.

While in Birmingham, Dennison was able to sell the old machinery to the Anglo-American Watch Company (that he may have had a hand in forming). It subsequently changed its name to the "English Watch Company." Dennison remained in England until his death on January 9, 1895, during which time he manufactured watch cases. (6)

There is no accurate count of production of the Tremont (Melrose) Watch Company. Crossman states that it "made a few thousand watches." Hauptman gives the opinion that "less than twelve thousand full plate Tremont and Melrose movements were made." (7) Shugart claims that the "Tremont Watch Company produced about 5,000 watches" and Melrose about 3,000. It is difficult to find any substantiation for any of these statements.

References

1. Edward Howard, "American Watches and Clocks," in Chauncey M. Depew (ed.), *One Hundred Years Of American Commerce,* (New York: D. O. Haynes, 1885,) p.542. Bailey states that Dennison left Waltham in 1862--a discrepancy of two years (see Chris Bailey, *Two Hundred Years of American Clocks and Watches* (Englewood Cliffs: Prentice-Hall, 1975), p. 252.

2. Wesley R. Hauptman, "America's Half-Breed Watch, NAWCC *Bulletin,* XII (Apr. 1966), p. 270.

3. Charles S. Crossman, *The Complete History of Watchmaking in America* (Adams Brown Reprint, 1885-1887), p. 106.

4. Hauptman, "America's Half-Breed Watch," p. 272.

5. Hauptman, p. 273.

6. Townsend states that Dennison returned to the Boston Watch Company in 1866 and sold the Melrose Watch Company to the "English Watch Company" in 1870. This statement does not agree with those of other researchers. See also Joseph Dean, "The Tremont Watch Company," NAWCC *Bulletin,* IX (Feb. 1960).

7. Hauptman, p. 275; Shugart, p. 240; Crossman, p. 107.

CHAPTER 8:
THE UNITED STATES WATCH
COMPANY OF WALTHAM

What became the United States Watch Company of Waltham, Massachusetts, began as the Waltham Watch Tool Company in 1879-- a small, private enterprise. The business was started and run by the Nutting brothers in the second story of a building on Crescent Street that was used for other purposes. The Nuttings were machinists who had been employed in a shop in that factory for several years. They decided that there was a better future making watch machinery and tools.

Enter Charles Vander Woerd

In 1882, the Nuttings were joined by Charles Vander Woerd, who resigned his employment at the Waltham Watch Company in the same year. Woerd, an immigrant from Leyden, Holland, was born on October 6, 1821. He was first employed in his father's business in the manufacture of telescopic and electrical equipment. He emigrated to the United States in 1844, and his first job here was in the Seth Adams machine shop in South Boston, Massachusetts. He remained there until 1853, when he moved to Cambridge. From there, he transferred to the machine shop of the Waltham Watch factory in 1857. (1)

Woerd distinguished himself as a machinist and inventor. He moved to the Nashua Company in 1859, when that company was established, and while there developed his machine for making epicycloidal forms on cutters for wheels and pinions. An epicycloid is a curve generated by the motion of a point on the circumference of a circle, the mathematical equation of which involves the use of sines and cosines. This machine removed the guess-work from making gears for watch trains.

When the Nashua Company was sold, Woerd returned to

Table 10

UNITED STATES WATCH COMPANY CHRONOLOGY

Date	Item
1884	The U.S. Watch Company organized with $50,000 capital by Charles Vander Woerd and the Nutting brothers, who transfer the Waltham Watch Tool Company name to the Hopkins Watch Tool Company
1884	Company produces first "Domed" model watches
1887	Vander Woerd leaves company; last dome watch produced
1888	New standard 16 size watch produced; production reaches 50 movements per day
1888-1896	Watches produced with other names on dials, including Suffolk and Columbia
1896	The United States Watch Company of Waltham dissolved as a Massachusetts corporation
1903	Theophilus Zerbrugg, of Marion, New Jersey purchases the Waltham property, which was leased to the Keystone Watch Case Company
1903-1910	High grade movements produced at the U.S. Watch Company plant for the E. Howard Watch Works (the Keystone Howards)

Waltham.   His next contribution to the technical aspects of watch
manufacturing was to improve jewel-making machinery.   For this, he
was given a cash award and a gold watch and he was made assistant
superintendent.   He invented other machines that facilitated the
manufacturing process, but his greatest development was the automatic
screw machine.   It combined three operations into one: turning the
outside diameter, cutting the screws and forming the point.   The
process was so simple that as Crossman said, "One girl will attend to
several (machines), each turning out 3,500 screws per day." (2)

Woerd's mechanical genius gained him the job of superintendent
of the entire Waltham Watch Company factory in 1875.   He held that
position until he left to join the Nuttings in 1882.   After a brief stay
in Europe, Woerd approached the partnership task in earnest.   He
convinced the Nuttings to concentrate on watch manufacturing, and he
outlined his plan to enlarge the operation substantially.   He enlisted
the financial support of E. C. Hammer, of Boston, who was the
Danish Consul.   They voted to go public.   The United States Watch
Company of Waltham was organized with $50,000 of capital.

The officers of the new company included: Thomas B. Eaton,
of Boston, President; Hammer, Treasurer; Chandler E. Edgecombe,
Woerd's son-in-law, Secretary.   Granville Nutting and Dennison
A. Wright also served on the board of directors.   The directors
attempted to name the company the "Waltham Watch Company," but
because of a suit filed by the American Watch Company the name
Waltham was forbidden. (3) Thus, the United States Watch Company
was born and the name of the Waltham Watch Tool Company was
transferred to the Hopkins Watch Tool Company, which moved into
the same building on Bedford Street.

Land was purchased in the northeast part of Waltham, and a 3-
story, brick building, 25 feet by 100 feet, was erected for
manufacturing Woerd's watch.   His first watch was 16-size, three-

quarter plate in three grades, and had a slow train. It also had a very large mainspring, which required that the plate above the barrel be raised to accomodate the center wheel. This "Domed" model required a special case, and for this reason was not popular. The model was discontinued in November 1887, the same year Woerd left. He moved to the San Diego Valley in California and died there on December 28, 1888. (4) Approximately 3,000 of the domed model were produced (about 300 days at 10 watches per day). By this time, the company had expended "upwards of $200,000 . . .but the results were far from satisfactory." (5)

FIGURE 20   *U. S. Watch Company "Dome" Model, number 2,904.*
It is 16 size, 7 jewels, and has a coin silver case (33,474)

In a change of marketing plans, the company offered a new, totally-redesigned movement, also 16-size. It was open face, had a standard case, a quick train and a conventional lever. These were made in nine grades. The new watches were more successful, allowing the company to expand both its capital and its capacity. It introduced an 18-size watch, as well as 14, 12, 6 and 0 sizes. The factory had 70 employees turning out about 50 watches per day. Shugart states that the total production of the company reached 802,000. (6) The top grade was the 18-size "President," which had gold settings, 17 jewels, and was adjusted to 5 positions.

FIGURE 21   *0 size, U. S. Watch Co. model, number 737071, with gold-filled Nauvoo case*

U. S. Watch Company Output

The output of the United States Watch Company includes movements with the company name and with other names on the dials. One such was the "Suffolk Watch Company." These watches were 0 size, 7 jewels, lever set and many had duplex movements.

Shugart claims that there are 18 size Suffolk watches, but the author has not seen any. Some watches were marked "Columbia." The Columbia Watch Company was founded by Edward A. Locke in 1896 and sold out to the Suffolk Watch Company in 1901, which, in turn, was sold to the United States Watch Company. Columbia manufactured 0 size, 4 jewel

FIGURE 22   *0 size, Suffolk Watch Co., Boston, Mass.,number 216,681.   The yellow gold filled hunter case is a 20 year Montauk, produced by Fahy's*

watches with duplex escapements. These open-faced gilt movements do not have serial numbers. They are engraved "Columbia/Watch/Co./Waltham,Mass./USA" on the plates. Columbia production is particularly confusing, because the company made watches in gilt for the Atlas Watch Co., New York and the Hollers Watch Co., Brooklyn. More confusing, other Columbias were ebauche. Figure 23 offers a comparison of a "true" Columbia made in Waltham and a nondescript version, made only God knows where. The real Columbia is 0 size, stem wind, duplex, with Roman dial and a yellow gold filled case.

In 1899, another brand of 0 size watch appeared--it was marked "Suffolk Watch Co./Boston, Mass." and was 7 jewel, nickel, hunting case. Known models have serial numbers of between 202,626 and 225,068. (7) In 1901, the Suffolk Watch Company succeeded the Columbia Watch Company and in a short time the new firm was

FIGURE 23   *Two Columbias Compared*

controlled by the Keystone Watch Case Company, which merged with the United States Watch Company.

In fact, the entire history of the United States Watch Company is confused. Bailey states that the company "failed in May of 1896, and in May of 1901 was purchased by some New Jersey backers who, in 1905 sold the factory to the Keystone Watch Case Company of Philadelphia." (8) Sanderson tells us that after some changes in management in 1895, the land and buildings were sold to the executors of the estate of E. C. Hammer and that in 1896 the United States Watch Company "was dissolved as a Massachusetts corporation." (9)

In May 1901, Theophilus Zerbrugg and others incorporated the United States Watch Company of Marion, New Jersey and bought the land and buildings of the Waltham company in 1903 to supplement the New Jersey plant. Zerbrugg leased these to his Keystone Watch Case Company. It was his intention to use the Waltham plant as an outlet for the Keystone high grade cases. The equipment and personnel of the Suffolk (Columbia) Watch Company were also transferred to the Charles Street factory. High-grade movements were produced there until 1910, when the plant functioned as the E. Howard Watch Works.

Summar states that after Keystone acquired the United States Watch Company in 1901, "most grades were discontinued." But the the Suffolk Watch Company machinery was moved into the United States Watch Company factory and a new 0 size movement was introduced in 1902. These were engraved "United States/Watch Co./Waltham,Mass./USA" on the barrel bridge and "A New/Watch Company/at Waltham/Est'd 1885" on the train bridge. Since later models were marked "New York" instead of Waltham, it is difficult to disentangle the serial numbers. Summar has seen numbers from 624,002 to 837,000. These 0 size watches were made with 7, 11, and 15 jewels, and the 15 jewel models are not marked. Some 7 and 11 jewel grades are marked. At the same time, the 0 size U. S. Watch movement was renamed the "Betsy Ross." Summar has information of movements in the 841,910 through 906,715 range. It is not certain when production of these 0 size watches ceased.

The Question arising from the above changes is what was produced at the United States Watch Company at Waltham after 1896. As we mentioned above, Suffolk and Columbia labels were part of the total output. It is clear, now, that the United States Watch Company continued to make watches after that date. Thomas L. DeFazio has a United States Watch of 16 size, 15 jewels, number 778,504 that is circa 1900. (10) And Henry Fried relates a conversation with a "Pop" Hoyt, in which he explains how he fitted, finished and timed United States Watches at the Howard plant in 1908. Presumably, this is the Keystone Howard of the post-1903 period. (11)

References

1. This brief history is based on Charles S. Crossman, *The Complete History of Watchmaking in America.* (Adams Brown reprint of Jewellers Circular and Horological Review, 1885-87), pp. 182-185.

2. Crossman, *Complete History,* p. 183.

3. Edmund L. Sanderson, *Waltham Industries* (Waltham: Waltham Historical Society, 1957), p. 105.

4. Chris Bailey, *Two Hundred Years of American Clocks and Watches* (Prentice-Hall, 1975), p. 205.

5. Crossman, *Complete History,* p. 185.

6. Cooksey Shugart, *The Complete Guide to American Pocket Watches* (Cleveland: Overstreet, 1981), p. 243. Once again the figures conflict. Criss puts the total production at "about 18,000 watches, all stem wind." (page 67)

7. Donald J. Summar, "Columbia, Suffolk, United States and Betsy Ross 0-Size Movements," NAWCC *Bulletin,* XXVI (Dec. 1984), p. 718.

8. Bailey, *Two Hundred Years,* p. 207.

9. Sanderson, *Waltham Industries,* p. 105.

10. "Answer Box," NAWCC *Bulletin,* February 1973, p. 1004.

11. "Answer Box," 'NAWCC *Bulletin,* October 1974, p. 621.

In Chapter 6, we described how Dennison and Howard established the first American watch manufacturing company-- The Boston Watch Company. Its life was quite brief; it produced watches from 1853 to 1857. When the company declared bankruptcy, Edward Howard returned to his Roxbury plant to start his own company. Charles Rice took much of the machinery of the Boston Watch Company to cover a loan he had made, and the remaining assets were purchased by Royal E. Robbins, of New York. He represented himself and the firm of Tracy & Baker, of Philadelphia, which had furnished the Boston Watch Company with cases. Robbins owned 85 percent of the new company. He kept Aaron Dennison on as superintendent.

Royal Robbins was the driving force behind the success of the Waltham Watch Company. A Connecticut Yankee, he was born in 1824, and at age 17, he visited an uncle, Chauncey Robbins, of Birmingham, England, who was an exporter of goods to the United States. Some of those goods were watches, and it was there that Royal made his first contact in the watch industry. When he returned to the United States five years later, he set up his own import business dealing in English watches. In 1848, he formed the company of Robbins Brothers, which included his brother David and Daniel F. Appleton.

Daniel Appleton, who was born in 1826, was a native of Marblehead, Massachusetts. His parents moved to Portland, Maine in 1853, where Daniel was educated and got his start in the watch business. He left Maine in 1847 to seek his fortune in New York. It was there that he commenced his long-time association with Royal Robbins, who was advertising for a clerk.

For Aaron Dennison, who became an employee of the new

Table 11

.THE AMERICAN (WALTHAM) WATCH COMPANY CHRONOLOGY

Date	Item
1857	Charles Rice took over some assets of Boston Watch Company; Royal E. Robbins purchases remainder
1857	Appleton, Tracy & Co. started as a partnership
1858	Name of company changed to American Watch Company
1859	Several leading employees of the American Watch Company leave to establish the Nashua Watch Company
1861	A. Dennison leaves Waltham
1874	The London office was opened with E. B. Dennison as sales agent
1876	Waltham production surpassed Swiss imports for first time
1883	Ezra C. Fitch, New York Sales Agent, becomes General Manager at Waltham; he becomes President in 1886
1885	Company name changed to American Waltham Watch Company
1892	The first railroad watch produced
1902	Royal E. Robbins dies; company taken over by his sons, Royal and R.C.
1906	Company name changed to Waltham Watch Company
1910	Royal and R.C. Robbins resign; E.Z. Fitch takes over
1911	Waltham diversifies by producing 8-day clocks and speedometers.
1912	Conover Fitch becomes Vice President and General Manager

Table 11 (Continued)

1915–1918	Waltham produces times fuses in World War I
1921	H.L. Brown replaces C. Fitch as General Manager; he is replaced by G.K. Simond as company losses mount
1923	F.C. Dumaine becomes President and Treasurer
1923	Company name changed to Waltham Watch & Clock Company
1924	Employees go on strike for first time
1925	Company name changed to Waltham Watch Company
1930	Tariff Act of 1930 increases duties on Swiss movements; they increase smuggling of movements
1931–33	Waltham begins to emphasize wrist watches
1933	Codes of Fair Competition in watch manufacturing established by the National Industrial Recovery Administration
1940	Waltham produces for war a third time
1944	Dumaine Sells out to Ira Guilden
1948	Waltham petitions court for reorganization; RFC grants $6 million loan
1949	John J. Hagerty assumes leadership of company
1950	RFC takes over company
1951–52	Waltham petitions President Truman to limit Swiss imports; request denied
1957	Company becomes the Waltham Precision Instrument Company

company, the bankruptcy of the Boston Watch Company was a great loss. As he wrote, "My aim is, if my health is as good as it always has been. . .to have a nice little establishment of my own on our side (of the Atlantic) to turn out in my own name as good a watch as can be made in any part of the world." (2)

Appleton, Tracy & Co.

Hauptman underscores the great importance of Appleton, Tracy: it "was indirectly resposible for the birth of the machine-made watch industry in the United States and perhaps in the world." (3) It produced more watches than all of its predecessors in the short period of 1857 to 1859. The companies that preceded it were mainly "changes of corporate names."

Appleton, Tracy & Co. began on a disagreeable note. When the articles of co-partnership were completed, Theodore W. Baker, who as an active partner, disagreed with their content. When Appleton and Tracy stood firm in their demands, Baker sold his share to Robbins. Thus, the new firm became Appleton, Tracy & Co.

Robbins succeeded Dennison as chief executive, and Dennison was sent to England to obtain dials, watch materials and skilled craftsmen. At this point, the technology for producing a good machine-made watch was known to Americans. What Robbins provided was the skill to market the output and to finance the growth of the company, that was to survive for a century. William H. Kieth, Treasurer of the land company who later became President, was added to the management staff. Finding skilled workers was a more difficult problem. Dennison could not induce the English to emigrate, but Moore claimed that this failure "was not crucial." (4)

Robbins and Dennison assembled a group of watchmaking "stars," including Nelson P. Stratton, who had helped the Pitkin Brothers in 1837; Charles S. Moseley; John J. Lynch; Charles

T. Parker and P. S. Bartlett.

Charles S. Moseley, who became foreman of the machine shop, was born in Westfield, Massachusetts on February 28, 1828. When he was 9 years old, his parents moved, briefly, to Illinois (a fact that probably influenced him later to move to the Elgin Watch Company). They returned to Westfield, and he was soon apprenticed in Hartford, Connecticut for six years for the machinist trade. After working in a rifle factory, Moseley was offered a job by Dennison, Howard & Davis.

Nelson P. Stratton, after completing his apprenticeship with the Pitkins, was employed at the Springfield Armory. He must have been dabbling in watch repair, because he was offered a job by Aaron Dennison in 1850. He was made Assistant Superintendent to Moseley. He was given the task of designing an 18 size movement. Later, he was sent to England to learn how to gild movements. He easily mastered that process, and he set up the electro process that was used by the company until he left to participate in the Nashua venture.

The depression of 1857 was a serious threat to the survival of the company. Robbins met the loss of revenue with a cut in wages and the employment of young women, and at the same time he paid premiums to highly-skilled workers. By cutting wages, he was able to keep his new factory operating and he accumulated an inventory of watches, most of which were sold later.

The early Appleton, Tracy watches were left over from the Dennison, Howard & Davis inventory. These were stamped with the Appleton, Tracy name on the movements and the name was put on the dials. Next, Robbins induced the American Institute of New York to award the company a medal for producing a superior watch. (5) Using all of his advertising skills, Robbins brought the award to the

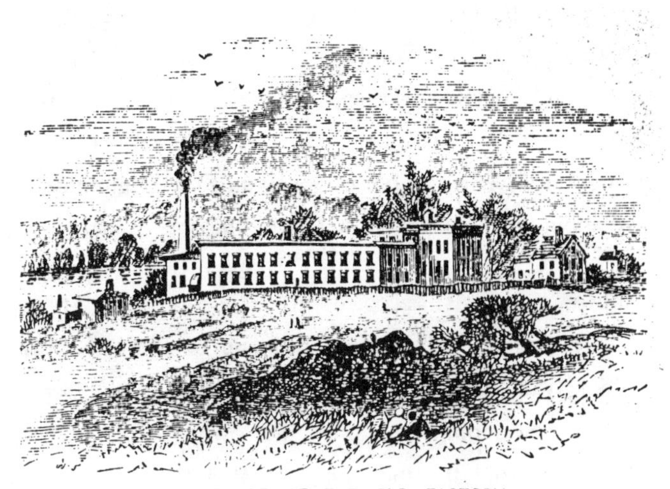

FIGURE 24 *The American Watch Company Factory, 1857*

attention of the public and the trade. At the same time, he reduced the price of some of his watches from $40 to $12 and sold them under the C. T. Parker label. Others were sold for $23.

In early 1858, the P. S. Bartlett movement was introduced, which was to become one of the most famous names in United States watch history. It was named after Patton S. Bartlett, a key member of the production team. It was the standard 1857 model, with 7 or 11 jewels and was priced at $15--$3 more than the C. T. Parker. The only difference between the two was some engraving on the balance cock. Presumably, Robbins was applying some modern sales techniques for selling his watches. Near the end of 1858, Appleton, Tracy produced the first American machine-made stop watch--the Chronodrometer. It was ingenious, but impractical, because all the hands stopped at one time, forcing the user to estimate the time.

The American Watch Company

Although sales picked up, Appleton, Tracy barely had enough

liquid capital to keep the company going. In the spring of 1858, Robbins proposed that the Waltham Improvement Company--a land company--take over Appleton, Tracy. Robbins' plan was to move the company closer to New York, near the sales office. (6) In May, the directors of the Improvement Company appointed a committee to confer with Robbins about purchasing the "Watch Factory Estate."

The above plan was accepted in June and the Massachusetts legislature was petitioned to change the name to the American Watch Company, which was accomplished in February 1859. Robbins was manager of the new company, since he owned 62 percent of the common stock. He sold the old company at a figure not much more than cost, but wisely, he accepted payment in shares of the American Watch Company. As Moore said, in this deal, he "cleaned up." (7) The merger also meant that the watch company now owned the real estate too (the old Bemis farm).

The officers of the American Watch Company were: Dr. Horatio Adams, President; William H. Kieth, clerk and Robbins, Treasurer and General Manager. They succeeded in increasing sales and profitability. The first dividend was paid in March 1859, and in the next quarter, 3,179 watches were produced. (8)

The Threat from Nashua

The early success of the American Watch Company spawned a competitive threat from a number of "breakaway" employees, who established a watch manufactory at Nashua, New Hampshire. (9) They were led by Belding D. Bingham, who was from Nashua, and who had owned a watch and jewelry business there. He made some clocks and regulators, but decided to move to Waltham to learn the watch business. It was there that he was associated with Nelson P. Stratton, the Assistant Superintendent, who was also interested in manufacturing a fine watch. They became good friends and talked always of producing jointly a precision timepiece. They visited Nashua in the fall of 1859, and shortly after became associated with L. Noyes, who provided the capital for the formation of the Nashua Watch Company. They chose Virgil C. Gilman, President; Thomas W. Lowell, Secretary and Noyes, Treasurer. (10)

Stratton was able to attract some of the best minds in the watch business to Nashua, including Ira G. Blake, James H. Gerry, Charles Moseley and Charles Vander Woerd, which was a severe loss to the American Watch Company. These people set about to design the finest timepiece ever made in the United States up to that time. They purchased the old Washington House and converted it into a factory. Woerd built the machinery, Gerry made the escapements, Charles Blake supervised the plate room and Josiah Morehouse, the dials. (11) Bingham, Moseley and Stratton did the actual designs.

The movements that were produced were 16 and 20 sizes, three-quarter plates, gilt, adjusted, with Stratton's patented barrel. They were particularly attractive for their top plates. Several technical improvements in watchmaking occurred at Nashua. For example, the isochronal balance wheel and a vibrating hairspring. Also, while at Nashua, Woerd invented machinery to grind gears in epicycloidal shapes.

The Nashua Company produced about 1,000 movements, but none went to market because of economic factors. In 1862, the company was short of funds. The original $53,000 was spent, and when a call went out to the shareholders to increase their investments, few responded. The Civil War reduced the demand for fine watches, forcing the owners to sell the inventory and the company to the American Watch Company. The $53,000 price was undoubtedly a great bargain. Stratton, Moseley, Vander Woerd and Gerry were reemployed by the Waltham company.

Charles W. Fogg, who was the Superintendent of the American Watch Company, was sent to Nashua to take charge of the factory there, but in the fall of 1862, the plant was moved to Waltham. In 1865, he perfected his "Fogg's Patent" for a threaded pinion. The pinion screwed into the center wheel so that when the main spring broke, it unthreaded itself. This reverse action meant that the barrel spun around and stopped, without damage to the teeth.

The Nashua Watch Company became "The Nashua Department," which was operated in a separate building until 1872. The original 1,000 movements were corrected and sold with new top plates that carried the Appleton, Tracy label. Stratton's vibrating hairspring was replaced with the Breguet (overcoil) spring. The 20 size model was discontinued in 1867.

The next achievement of the Nashua Department was the introduction of the 18-size "Crescent Street" for the railroad market. It was modelled by Woerd and first sold in 1869. It was discontinued in 1874, after 17,500 were produced, probably because it required a special case--as did most of Woerd's watches. (12) The Crescent Street name was revived in 1883 as a top-grade, full-plate watch. It is also notable for being the first stem wind watch produced by Waltham (650,001 to 651,000). A few experimental stem wind watches may have been produced before this run, but they were adaptations, using the Abbott pattern winding device.

Elsewhere, we have called this "Crescent Street" watch "The Elusive Model A." This 18 size "Crescent Street" was named after a street aside the factory. As we stated above, it was introduced as a railroad watch, but the date of first sale and the existence or non-existence of two models is a matter of controversy. Some descriptions of Waltham watch movements list a Series A and a Series B. No date is given for the Series A, but the Series B is referred to as an 1870 model. It is full plate, 15 jewels and has all the improvements available at that time: garnet jewels, patented pinion and compensated balance.

Townsend presents a diagram of movement number 500,516, which is a Model B, and he presents the production runs for the two models, indicating that the Model A is considerably scarcer than the Model B--2.500 compared to 15.000. (13) Ehrhardt gives the serial numbers for the two models, and he offers a diagram of movement number 690,046--which is also a Model B. (14) Hauptman may be correct when he states, " My conclusion is that they were designed and a prototype was built late in 1869; that a quantity was produced late in 1870 but none were sold until just after the first of the year 1871." (15)

The issue is not settled. Willard Halsted, in a letter, wrote, There definitely *is* a difference between the Waltham 18 size Series A and Series B." He refers us to the 1885 Waltham Catalog (pages 14 and 22) where he states we will find a space milled out for the center wheel on the series A but not on the Series B model. He also notes that several parts on the two models have different numbers (ie main springs 2201 and 2202). But Harrold, also in a letter, states, "I agree that there is no difference between the two series, and believe that the distinction was an accident, created when the Waltham serial number list was completed, circa 1940." (16)

Figure 25 illustrates an 1870 B Crescent Street model. It is number 500,566, with 17 jewels and a Dueber Silverine case (1,504,910).

The final technological achievement of the Nashua Department was the elegant 1872 model. About 60,000 of these three-quarter plate movements were made with "Amn Watch Co." and "American Watch Co." on the movements. These watches achieved world recognition. They were exported, and they were honored at the

FIGURE 25   *1870 B Crescent Street model*

American Centennial Exposition in Philadelphia in 1876, where the American Watch Company had a large exhibition. The company won four first prizes and as Cuss stated, "The Philadelphia Exhibition of 1876 greatly impressed the Swiss; their methods of production were reorganized as a consequence." (17)

The Golden Age

Moore has called the years of the Robbins management at Waltham "The Golden Age." (18) Besides assembling a remarkable group of watchmakers, Robbins had the advantage of very favorable economic trends. Between 1860 and 1880, the United States population grew by 11,500,000 and per-capita wealth rose by 70 percent. In 1860, the American Watch company paid a 5 percent dividend--the first for any American company, and dividends were paid annually. Abbott tells us that Robbins earned $337,000 in 1865, the largest income in Boston. (19)

Robbins' success was related to his ability to oversea

development of new machines and processes to manufacture parts, at the same time he was simplifying other mechanisms. Many parts were eliminated from the Waltham watches. By the time of the Centennial Exhibition, "fully three-fourths of the parts which at that time were employed by English watchmakers had disappeared entirely from American watch construction." (20) By this approach, Robbins brought "order out of chaos." The goal became to produce simple, accurate and reliable movements. But until the end of the 1860's, the manufacture of dials, jewels and hands were strictly European skills. Robbins turned his attention to not only making these parts in the United States, but making them by machine. And, the entire process was performed under one roof, so that, "every part of the watch is cut in their establishment by aid of machinery, graduated to microscopic exactness and working with a delicacy of touch that the fingers would strive in vain to emulate." (21)

The year 1861 was a noteworthy one for the new company for two reasons: it marked the end of Dennison's association and Robbins was married. Dennison and Robbins never became friends, but in spite of their bickering, Dennison remained as superintendent. Robbins found Waltham dull (compared to New York) and sought the attention of Mary E. Horton. His search was successful, and they were married on October 21, 1861. They left almost immediately for a honeymoon in Europe . During their absence of three months, William Keith was put in charge of operations. There developed a rift between Keith, acting on orders from Robbins, and Dennison. It was not a frivolous matter, because it involved the market for a soldiers watch. Dennison was convinced that there was a very large market for an inexpensive watch for soldiers fighting in the Civil War. Keith was not interested particularly because of the depression and because the company had already designed two inexpensive watches: the "C. T. Parker" and the "P. S. Bartlett." Upon Robbins return, he forced a vote in the Board of Directors to abrogate Dennison's contract. Dennison was correct in his assessment of the demand for a "soldier's watch," but Robbins stuck to his belief that such a watch should be manufactured without a large investment in tools, dies and jigs.

It is difficult to unravel the true history of the low grade movements at the American Watch Company. It began with the production of the "C. T. Parker" movement, named after Charles T. Parker, a specialist on the technical staff, in November 1857 (serial numbers 8101-8300). These watches were made in 7 and 11 jewels, and according to William Taff, the highest number was 11,998. (22)

For no obvious reason, the name of the movement was changed to J. Watson, which produces another horological puzzle. The question is , who was J. Watson? Hauptman could find noone with any connection to Waltham with that name, but he did uncover a John Watson in the Boston Business Directory for 1850. The J. Watson movements were produced in two runs: 23,601 to 24,300 and 28,201 to 28,700. This is a curious aspect of the American Watch Company production. If this "J. Watson" was to be the "soldiers watch," why did most of them have London on the barrel bridge and why were most sold in England?

The Watson movements were then renamed "Wm Ellery," after one of the signers of the Declaration of Independence, and this watch became known as the "Soldiers Watch"--the watch over which Dennison was fired from the American Watch Company.

The basic three-quarter plate movement was introduced in 1861 in 7 and 11 jewels, but it was discontinued in 1868. The next new watch was made for women. It was made public in September 1861 as a 10-size, three-quarter plate, Key wind, key set. Two models were offered: a 13 jewel "P. S. Bartlett" and a 15 jewel "Appleton,

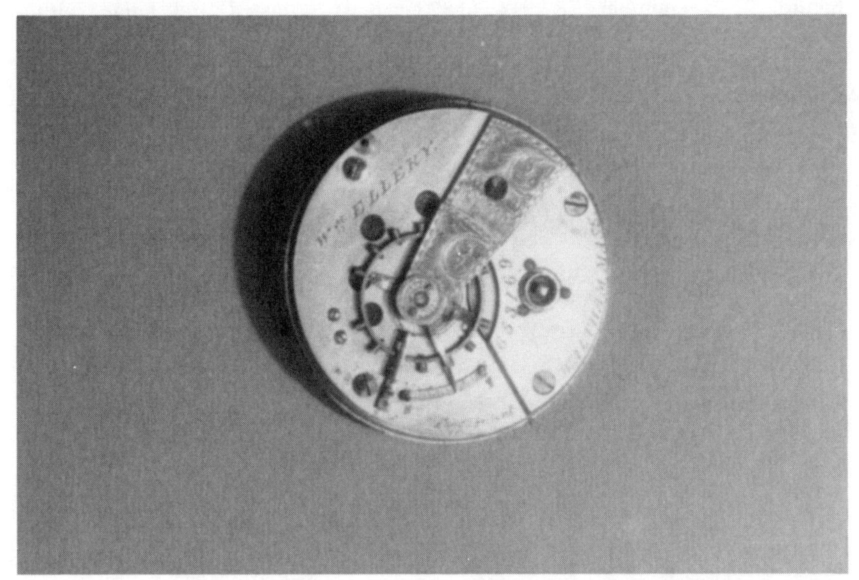

FIGURE 26   *An example of the 1857 William Ellery model,
used earlier as the soldier's watch.   It is 18 size, 11 jewels,
number 653,169 (1872), with Fogg's Patent*

Tracy & Co."   When the Nashua Department was established at
Waltham, the three-quarter plate 20 and 16 size movements replaced
the 18 size as high-grade models.

The American Watch Company utilized several inventions and
innovations, only some of which benefited the company.   The Fogg
vibrating hairspring, for example, patented on January 2, 1864, was
soon discarded.   But Fogg's patented pinion was more useful.   Its
threaded main wheel helped to ease the jolt of a broken main spring.

While most watches of the American Watch Company were very
high grade, the company did produce an inexpensive model-- the
Home Watch Company.   These watches were 20 1/2 ligne (over 18
size), 17 jewel, kew wind, key set, with jewelled lever escapement.
They were the standard Model 57, and they were produced from
approximately 1872 to the end of the century.   The example shown in

Figure 27 is number 782,914 (1873); it is adjusted and came with a yellow gold filled hunter case, GWL 29649 (George W. Ladd, casemaker).

FIGURE 27   *A Home Watch Co. Model*

In producing these Home watches, Waltham engraved them with "Boston" instead of "Waltham" to conceal their origin and to disassociate them from the higher quality American Watch Company, Appleton Tracy & Co. and P. S. Bartlett models.

After the Home watch production was terminated, Waltham renewed its attempt (in 1909) to capture the low-priced market with the introduction of the Equity Watch Company model, also engraved "Boston." The first production run was 17,874,001- 9000 (1909); the last was 22,401,000-5000 (1918). Most were 1908 model, 16 or 16 1/2 size, 7 jewels, open face and adjusted. But there were also 17,400 12 size and 6,000 6/0 size produced. Some of these were hunting cased.

The watch shown in Figure 28 is an oversize 16 1/2, 7 jewel,

Equity, with serial number 21,238,744 (1918), with a Comrade 20 year gold filled case (number 4,097,337). The early Equity line of watches has produced one puzzler: Some of them had a numbering separate from the regular Waltham numbering. Larry Treiman discussed one of these watches; it has serial no 2880. (23)

FIGURE 28 *The Equity Watch Co., Boston*

Marketing Efforts

As we stated above, Robbins pioneered the advertising of watches in periodicals and at exhibitions. He usually combined the showing of machinery and watches at one time. He also used newspaper advertising very effectively. In 1865, total advertising expenditures were $7.3 million; by 1872, these had grown to $63.6 million, after which they declined precipitously in the depression of 1873. (24) These ads had a positive effect on sales and profits. In 1866, a cash dividend of 150 percent was declared, and a new subcription of common stock was offered, which increased the total capital from $300,000 to $800,000. (25)

Robbins also thought much of establishing a world market for his watches. Even during his honeymoon, he attempted and failed to secure a London agent. During the panic of 1873, Robbins shut down the Waltham plant and worked on designs for movements for the foreign trade. The annual report for 1874 wrote of the need "to make goods especially suited to the tastes of foreign markets." (26) Thus, a London office was opened as a selling agent for Robbins & Appleton. Nelson Stratton was sent to London, during which time, he made contact with Aaron Dennison in Birmingham. The result of these negotiations was the employment of Aaron's son, Edward B. Dennison, as a sales agent. The London move was a great success. Total sales rose from 30.6 thousand pounds in 1876 to 107.2 thousand in 1883.

FIGURE 29 *A Waltham Advertisement*

Although Dennison was not employed by the American Watch

Company, he played a part in its success. He commenced manufacturing watch cases using the same methods he had developed at Waltham. These cases were high quality and were a complement to the high-quality movements to which they were combined. Dennison was able to supply all the silver cases needed by Waltham in England.

At the same time, the American Watch Company introduced three 14-size movements, which were very popular and which were produced for many years. The first three--the "Adams Street," "Crescent Gardens" and "American Watch Company" were three-quarter plate, 7,11, and 15 jewels. They were first advertised as watches for young boys and gentlemen. These original names were discontinued in the 1870's, but the 14 size was produced for many years. The later ones were stem wind, three-quarter or full plate and carried the names of "Bond Street," "Riverside" and "Hillside."

FIGURE 30 *The Basic 14 Size Model of the The American Watch Company. This Bond Street has 7 jewels and is pin set for the export market. It has an English-style, Silver case*

Meeting Competition

The introduction of new movements came at a poor time. The 1873 downturn in business was called the Great Depression of the nineteenth century, and according to the *Fitchburg Daily Sentinel*, in 1875, the company "stopped work for 5 weeks in all its departments except the machine shop." (27) But by 1876, the Waltham production surpassed Swiss imports for the first time. It was a very dramatic turnabout, undoubtedly linked to the favorable publicity of the Centennial Exhibition.At the Philadelphia Exposition of 1876 the Waltham Watch Company made its biggest sales pitch of the century. To quote J. E. Coleman, "[It] literally transported a whole watch factory for its exhibit and set up shop, making complete watches by machinery for all to see. It was here that our foreign watchmakers were first really impressed with machine made timepieces, and in no time the Swiss were installing watchmaking machinery." (28)

Swiss imports, which were 134,000 in 1875, declined to 75,000 in 1876, a 44 percent decline, while Waltham production grew from 49,243 to 84,737 in the same time. (29) The long-time Swiss supremacy in watch manufacturing had switched to the United States.

The sharp increase in production induced the directors of Waltham to expand the plant. In 1878, wings were added and departments expanded. This was accomplished in the face of a growing competition. Since most of the new companies were begun by Waltham alumni, it almost seemed that the company was competing with itself. Running down the list, we find the Nashua Company, started in 1859, with N. P. Stratton, Charles S. Moseley and D. B. Bingham as organizers. In the following year, Dennison left the American Watch Company to produce watches for the Tremont Watch Company. In 1863, James H. Gerry took some Waltham employees with him and founded the United States Watch Company of Marion, New Jersey, and in the following year, N. P. Stratton again was involved in starting a competing company--the Elgin Watch Company. The personnel were similar: besides Stratton, there were Charles S. Moseley, P. S. Bartlett, J. K. Bigelow, George Hunter and Otis Hoyt. The story was repeated stil another time in 1869, when J. C. Adams, J. K. Bigelow, Otis Hoyt, C. E. Mason and John Nickerson, all graduates of Waltham, started the Springfield (Illinois) Watch Company. Between 1869 and 1885, the Rockford, Fitchburg, Columbus,Fredonia, New Haven, Manhattan, Aurora and New York Standard Watch Companies were organized--most with Waltham personnel. (30) By 1890, there were 36 attempts at watch manufacturing in the United States, 19 having failed. At that time, there were 17 companies turning out 7,500 watches per day with 12,000 employees (5,000 female). There were also 47 watch case manufacturers turning out 6,000 cases per day, employing 5,000 persons. (31) The difference in the production of movements versus cases (about 1,500) went mostly to foreign markets.

The Waltham company continued to meet competition with innovations. Nearly all major technological changes in the watch industry were developed in Waltham. The Woerd "improved compensation balance" was just one example. Although it was introduced officially to the public at the Paris Exposition of 1878, it had been tried on the 1872, 16-size, three-quarter plate, 17-jewel watch (which actually had 19 jewels). Vander Woerd was granted his patent on May 2, 1878. His balance used copper-zinc instead of brass to control temperature. By use of intermeshing teeth, he was able to reduce the gain or loss in seconds due to temperature changes to under 2 seconds per day. (32)

It should be noted that it was in the Nashua Department that Woerd invented several new devices and processes. When this department was established in 1864, Woerd became its assistant superintendent. It was in this capacity that he designed the 1870, 18-

size, Series B "Crescent Street" movement. In 1876, he was made general superintendent of the entire Waltham production facility.

These technological improvements in the production of machine-made watches certainly made the Europeans, especially the Swiss, take notice. They sent Edward F. Perret to find out why Swiss imports had declined so precipitiously. After seeing the American display at the 1876 Centennial Exhibition, Perret, one of the Swiss Commissioners of the International Jury on Watches, noted that "American machine-made watches with interchangeable parts, produced at low and medium prices, which were more elaborate and better looking and which compared favorably in time keeping qualities with the costlier foreign made ones, especially those of Swiss make, would in a short time establish the United States as the leading watch producing nation of the world." (33) The prediction proved true. So true, in fact, that the Swiss remedied the defects in their watches by adopting American techniques.

One of Perret's statements is worth repeating here. He was surprised at the productivity and the quality he found, and he related a personal experience: (34)

> I asked from the manager of the Waltham Company a watch of a certain quality. He opened before me a big chest. I picked out a watch at random and fixed it to my chain. The manager asked me to leave the watch with them for three or four days that they might regulate it. 'On the Contrary,' I said to him, 'I want to keep it just as it is to get an exact idea of your workmanship.' On arriving at Locle I showed this watch to one of our first adjusters. . .who took it apart. At the end of several days he came to me and said, literally, 'I am astonished, the result is incredible. You do not find a watch to compare with that in 50,000 of our own make.' This watch, I repeat to you, gentlemen, I myself took offhand from a large number, as I have said. One can understand by this example how it is that an American watch should be preferred to a Swiss

watch.

In 1883, Ezra C. Fitch was brought to Waltham from New York City to take over as General Manager. He had been sales manager of Robbins & Appleton. Fitch was born in Germany, but was educated in the American public schools in Worcester, Massachusetts, where he learned the trade of watchmaking. He worked for Bigelow & Kennard, Boston watchmakers, before going to Robbins & Appleton.

FIGURE 31 *Two 6 size AWC 1873 Models.*
On the left is a 7 jewel Seaside,
number 2,797,311 (1885); on the right, a Royal,
number 2,767,437 (1884)

Fitch's job was to ward off the effects of the increased competition by offering new approaches to merchandising: producing watches to fill precise demands in the market place (style, shape, jewels, case, etc.). This new approach coincided with the development of department stores in major cities (Macy's, Gimbels, Wanamaker's, Jordan Marsh) and with mail order selling in rural areas (Sears Roebuck). The new marketing approach is reflected in the number of grades offered by Waltham: 21 in 1881 and 45 in 1896.

Fitch became President of Waltham in 1886, and Robbins remained with the company as Treasurer. Fitch also assumed the position of Assistant Treasurer, to relieve Robbins of the heavy burdens of that office. When Fitch returned to Waltham, the locus of power shifted . New York City was always considered the main office, with Boston and London as branches. The New York office, which began on Maiden Lane, moved to 182 Broadway, then to the "Waltham Building" on Bond street in 1871. When this was destroyed by fire in 1877, the company moved to new quarters again.

The intensity of competition in American industry led John D. Rockefeller and J. P. Morgan to develop combinations to "rationalize" their industries. The watchmakers responded also. In 1885, they created the National Association of Jobbers in American Watches, under which all manufacturers sold to jobbers at the same price. This form of cartel seems to have had little impact, and it collapsed in 1890, after which price competition became more ferocious, with companies cutting prices as much as 20 percent. In 1885, the company changed its name to the American Waltham Watch Company, perhaps to emphasize its national character in a competitive market.

Fitch presided over several important production changes. Although Howard had introduced stem-wind watches in 1869, Waltham waited until 1888 to produce such a device. But the major change in watch design was the reduction in weight and size. The full plate, 18- size watches were always considered a "man's" watch-- what we would call today "macho." These full plates were replaced with split plates--first the three-quarter plate, then the bridge models. When the popular demand shifted to 16 size, the full plate was abandoned. In 1890, Waltham dropped the three-quarter plate in favor of split plates. By separating the balance and the train under separate bridges,

Waltham was able to improve the grades of watches and to meet the railroad standards set in 1892. In addition, the split plates were far easier to repair, an important consideration since wages in the United States were considerably higher than those in Europe. Between 1890 and 1920, split plates came to dominate American watch movements. (35)

Joel Gross, of the Washington Chapter of the National Association of Watch and Clock Collectors, an authority on the Waltham Watch Company, calls the 1892 model "the second most interesting Waltham," the first being the 1857 model. The 1892 movement was Waltham's answer to the competition, particular from Elgin, which by that year produced nearly as many watches as Waltham. It was a time when the Jobbers Association collapsed and Waltham was compelled, once again, to cut wages and prices. But the 1892 model was 18 size and full plate--in effect, going against the trend of reduction of size. It was sold in 17-23 jewels, with the Vanguard as the top grade.

Legal Problems

As we stated in Chapter 8, when Charles Vander Woerd left the Waltham Watch Company, he convinced the Nutting Brothers to establish the United States Watch Company of Waltham. It appears that he wanted to capitalize on the the Waltham name, and, in fact, he advertised some of his watches as "Waltham watches." In 1890, the Waltham Watch Company went to court and obtained a preliminary injunction against the practice. The lower state court found that the United States Watch Company practice "led to consumer confusion and were an example of deceptive advertising." (36) The court ordered the United States Watch Company to cease using the Waltham name on its plates and dials unless it clearly distinguished its watches from those produced by the Waltham Watch Company. The United States Watch Company felt that this prohibition was too harsh, so it appealed this decision to the higher court. Eventually, the case went to the Supreme Court of Massachusetts.

The case, which was heard by Chief Justice Oliver W. Holmes in 1899, was unique, because it concerned whether the name of a city could be the subject of a trademark. The law of that time did not allow for a trademark of a geographical area. Justice Holmes, however, made an exception. He stated that in this case the name of the city was so associated with the name of the watch company that the general public asssociated Waltham with pocket watches. Thus, Holmes agreed with the lower court: that the United States Watch Company had to take every precaution not to confuse or deceive the public about its watches and that these had to be carefully distinguished from those of the American Waltham Watch Company. (37)

The second legal problem of the Waltham Watch Company involved the antitrust suits of 1903-07. The Antitrust Division of the Justice Department charged that the manufacturers of watches and cases had formed a watch trust, that they were selling watches overseas at prices less than those at home, that they were colluding in the setting of prices and that they were using pressure to force railroad employees to buy their watches. In the 1903 case, a retail dealer stated, "It has been no secret all along that you can't get an Elgin or a Waltham movement without its being certain that you are going to have a Crescent or a Keystone case. (38)

To understand this case, one needs to review developments in the United States before and after the election of 1896. In 1882, the Standard Oil Trust Agreement was signed, giving the company general supervisory power over all companies that were signatory to the agreement. In the rest of the decade, many trusts were formed, leading both major political parties to favor an antitrust law. The Sherman Antitrust Act was passed in 1890. This was just one product of much larger struggle. Democrats of the south and west (the agricultural, mining and working class interests) were trying to obtain political control from the financial and industrial interests of the north and east. (39) This struggle was linked to the downward trend of prices from 1873 to 1896. When William J. Bryan was defeated in 1896, it placed control of industry in the hands of conservative Republicans. They passed the Dingley tariff, which placed a 25 percent duty on movements and a 40 percent duty on cases. This meant that the companies that were colluding were protected from foreign competition. The watch companies, which were charged with monopolising output, were not tried in court. These were times of unregulated industry, and even the United States Supreme Court tended to decide cases on the side of property. President Roosevelt attacked the leaders of finance capitalism, but the Department of Justice decided that it did not have enough evidence for an indictment. The case was really tried in the nation's press, however, where the watch industry was a target.

The New Prosperity

The election of William McKinley in 1896 signalled the start of a new, long-term prosperity. By 1900, the United States volume of foreign trade reached $2.3 billion, with a favorable balance of $649 million--the largest in United States history. The Waltham Watch Company shared in this new prosperity. Its profits quadrupled between 1896 and 1901. But in the next year, Royal Robbins died, and his death set off a battle for control of the company between Fitch and Robbins' heirs. During the struggle, the company's marketing and sales programs deteriorated, as consumers expected and did not get their accustomed innovations in watches.

But conditions changed abruptly. First, the company name was changed in 1906 as the word "American" was removed from the title.

FIGURE 32   *An 0 size, 17 jewel Riverside, Model 1900, number 13,012,967 (1904)*

Then, the Premier Maximus was introduced in 1908, and finally, the Robbins family sold their interest in the company in 1909. This also meant the cancellation of the exclusive distribution rights, which Robbins & Appleton had claimed from the time Robbins became associated with the company.

The Maximus movements were the finest ever produced by the Waltham Watch Company. What a brilliant choice of name for a watch that Autry called "Waltham's Magical Maximus. (40) These movements were produced for about 40 years in a dozen models from 3/0 to 16 sizes. The Premier was introduced in 1908 and the Riverside in 1910. As an example of snob appeal, Waltham raised their prices to increase the prestige of these watches. The Premier Maximus was increased from $250 to $400 in 1914 and to $732 in the 1924 catalogue.

Although the Maximus models had a long production history, the

Premier and the Riverside models are very scarce. There has been much written as to which is the scarcest of the Maximus models. In 1980, Autry wrote that there were more Riverside models produced than Premier, but Lindberg found a typographical error in Autry's article. Lindberg's totals show 1,200 Premier and only 500 Riverside models of the Maximus. As he said, "The Riverside Maximus is widely regarded as the pinnacle of quality and grade of all Walthams. . .(it) stands out in quality and workmanship as the ultimate watch produced in Waltham and perhaps in America. (41) He also states that the 21 jewel is more rare than the 23 jewel model, contrary to what you find for other models.

What was this "Magical Maximus?" It was usually 16 size, open face, pendant set, with ruby jewels, double sunk dial, a gold train, double roller, micrometer regulator and a compensated balance. The dial showed the word Waltham in script. This beautiful watch was not alone, however. It had to compete with the Howard Keystone model, the Gruen Anniversary model and the Hamilton Masterpiece line--all in 23 jewels.

But competition from Switzerland intensified. Swiss exports were the highest in history, in spite of the high tariffs imposed on them. The Waltham management met the growing competition by diversifying. In 1911, they introduced the 8-day clocks; in 1915, the banjo clocks and time fuses; in 1916, automobile speedometers and blood pressure guages; and in 1917, phonograph motors. (42) Waltham's venture into clock manufacturing began with the purchase of the Waltham Clock Company, which was making clocks based on the English Elliot pattern. These high-quality clocks were produced until the 1930's.

As a consequence of its diversification efforts, Waltham was producing at full capacity, and the company opened four new plants: at Quincy, Newburyport, Gardner and Greenfield, Massachusetts.

FIGURE 33 *The 16 size, 23 jewel Riverside Maximus, number 10,562,616, model 1899. It is fitted with a Keystone gold-filled hunting case. It should be noted that this watch is misnumbered in the Ehrhardt Waltham volume, which records this serial number as an 1890 Lady Waltham*

Once again, Waltham was involved in war production in its new department Number 900, where time fuses were produced. The company hoped to duplicate its success of the Civil War, and it was only partially successful.

The prosperity of World War I (1914-1920) was propelled by a wave of new technologies, led by the development of the Ford moving assembly line. These techniques increased efficiency in manufacturing by reducing costs substantially. Many spoke of a new era of prosperity, and some began to call the 1920's "The Prosperity Decade." The primary postwar depression was sobering. Waltham, along with most manufacturing companies, found itself with a considerable inventory of unsold goods. To add to the financial burden, the company had invested in much new, expensive equipment.

Instead of a new prosperity, Waltham faced the brink of bankruptcy, as the demand for watches fell 35 percent in 1921. Ezra Fitch, who dominated Waltham for nearly 40 years, tried to hang on, but his management became more feeble, and eventually he became a victim of what Berle and Means called the divorce of ownership and control. (43) In fact, Fitch began to lose control in 1912, when in a reorganization his son, Conover, was made Vice President and General Manager. World War I postponed the inevitable, but in 1921, Harry L. Brown, the Treasurer was made General Manager, to replace Conover Fitch, who was in poor health. In 1921, Waltham owed $8 million, the inventory was overvalued, and the company was unable to compete in a market that showed a rising demand for wrist watches. In the face of these dismal economic conditions the banks stepped in to take over Waltham. (44)

Enter Frederic C.Dumaine

In 1921, the First National Bank of Boston, acting for all the creditor banks, took over control of the Waltham Watch Company. These banks decided that the Waltham management was incompetent, so they brought in Gilford K. Simonds, the respected head of the Simonds Saw and Steel Company, of Fitchburg. Under his driving leadership, the Simonds company had prospered. It was expected that he would duplicate this success at Waltham. But he found severe morale and disciplinary problems at the watch conmpany. He attempted to reorganize the company and to speed up production. But the long-lingering personnel problems reappeared. The Waltham departments were controlled by inefficient employees. Simonds sought to root out inefficiency, with some success. But it was not enough to satisfy the investment bankers. They asked Kidder, Peabody & Company to reorganize Waltham.

The investment banking firm chose Frederic C. Dumaine as Waltham's new chief executive. He contributed $125,000, for which he received 12,500 shares of common stock. Ezra Fitch became Vice President, but he performed only advisory functions.

Dumaine had come to Dedham, Massachusetts at a very early age and he attended public schools there. When he was only 14 years of age, he began working at the Boston office of AMOSKEAG. From there, he went to the company's mills in Manchester, New Hampshire, where he learned cotton manufacturing so well that he was returned to Boston as head of that office. While in Boston, he mingled with finance capitalists of that city, and he began to make very profitable investments for his own account.

Since Dumaine had no previous watch experience, he appointed Israel E. Boucher as General Manager. Together they ran the Waltham Watch Company for more than 20 years, but Dumaine was the unquestioned leader of the Board of Directors. He continued Simonds policies of trying to improve discipline and productivity at the plants. But, unfortunately, he tried to achieve these results by economizing. He cut research expenditures, cut the advertising budget and reduced the sales force. Then he cut wages in the recession of 1924, by as much as 40 percent in some departments. It was not an unusual move, since wages were reduced throughout the country, but the average 10 percent reduction at Waltham caused a deep resentment, that plagued the company for years, and it brought about the first strike in the history of the watch company.

The company won the strike after five months, but it was a misdirected effort, because it was attacking the wrong problems. During World War I, Waltham chose to emphasize war production (a mistake it repeated in World War II), but Elgin and Hamilton, its chief competitors, continued to produce and to improve watches, and the importation of Swiss watches grew alarmingly.

Dumaine's Production Policies

Dumaine believed that if Waltham could produce good quality watches that sales would increase. But to do this, he chose to reduce the number of grades from 180 (the number set by Fitch) to 37. (45) At the same time, Dumaine set out to put up to 70 percent of total production into wrist watches, and he increased the output of 8-day movements. (46) The original 8-day clocks were produced as 1907 and 1910 models and were 37 size. In 1924, a new model was designed that went on the market in 1926. This increase in clock activity was consistent with the name change voted by the company: in 1923, it became the Waltham Watch & Clock Company. The 1926 8-day model was produced until World War II, when a 22 size movement was added. These 8-day clocks were manufactured until 1953 in a great variety of movements, which included chronometers, aircraft clocks, watchman's clocks, industrial clocks, military clocks, automobile clocks and parlor clocks. They had 7, 9, 11 and 15 jewels. Hawkins estimated that a total of 1,801,700 8-day movements was produced. (47)

FIGURE 34 *An 8 Day Waltham Clock in a Mahogany Case*

WALTHAM
COLONIAL WATCHES

1929 C. A. KIGER

$100.00

No. 100-9   **No. 100-10**
Green Gold   White Gold
Butler Finish — Octagon Case
Price $107.00

Royal 19 Jewels, Temperature Adjusted and 3 Positions. 14K Green or White Octagon Case, Engraved Bezels, Brushed Back. Sterling Silver, 18K Applied Gold Figures.
Dial No. 473

Waltham Colonial "A" Watch No. 175 in Presentation Box

$80.00

No. 100-7   **No. 100-8**
Green Gold   White Gold
Border Engraved Case
Price $107.00

Royal 19 Jewels, Temperature Adjusted and 3 Positions. 14K Green or White Gold Case, Butler Finish or Border Engraved. Sterling Silve, 18K Gold Applied Figures.
Dial No. 424

Waltham Colonial Watches are
Cased and Timed at the Factory.
Each is supplied in a handsome
Presentation Box.

$75.00

No. 75R-13   **No. 75R-14**
Green Gold   White Gold
Engraved Center Case
Line Finish Back
Price $85.90

Royal 19 Jewels, Temperature Adjusted and 3 Positions. 14K Green or White Gold Case, Engraved Center, Burnished Bezels.
Dial No. 727

No. 60R-15   **No. 60R-16**
Green Gold   White Gold
Butler Finish — Cushion Case
Price $64.10

Royal 19 Jewels, Temperature Adjusted and 3 Positions. 14K Filled Green or White Cushion Case, Engraved Bezels.
Dial No. 1407

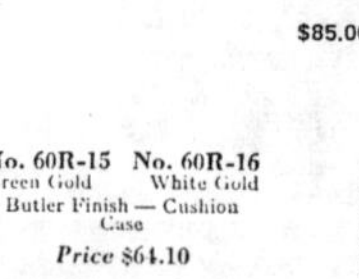

$85.00

No. 75-9   **No. 75-10**
Green Gold   White Gold
Butler Finish — Octagon Case
Price $85.90

17 Jewels, Temperature Adjusted. 14K Green or White Octagon Case, Engraved Bezels, Brushed Back.
Dial No. 1105

$55.00

Illustrations seven-eighths actual size

In the 1920's, the Waltham directors voted also to expand greatly the production of the "Colonial" series. Between 1904 and 1924, 161,985 Colonial movements were turned out in a variety of grades. The great majority were 14 size and the remaining runs were of 10 size. (48) Many were of the "Export" grade (1894 model, 7 jewels and unadjusted). A large number were Series 1420 and 1425 in 15-17 jewels. The first Maximus in the Colonial Series was produced in 1915. It had 23 jewels. Other Maximus models had only 21 jewels. The Colonial A model was very thin and came in a unique case size.

Figure 36 offers a comparison of two Colonial models--one old and one new. On the left is a 12 size, 19 jewel Riverside, number 14,188,284 (1905); on the right is a 10 size, 1924 model Colonial, number 29,164,716 (1937). It is Grade 217, 17 jewels with a Star 14K yellow gold filled case (2,224,992)]

FIGURE 36   *Two Colonials Compared*

All of the post-1924 Colonials (the "B" and "R" Series) were

12 size.   Most were produced between 1924 and 1929 in the "Prosperity Decade."   They were 17 and 19 jewel adjusted to 3 to 5 positions.

With the cost-reduction program in place and expansion of popular models, Waltham showed a profit in 1926 of $1.3 million--the first profit since 1919.   But the directors of the company could not be complacent, because there were some ominous signs all around them. In particular, some competitors were assembling Swiss watch parts in the United States.   It was not the first time that foreign parts had been used in the United States, because Dennison had produced his "Half-Breed" watch for the Tremont Watch Company.   But this was different; it was all-foreign movements that were being assembled here. As foreign producers came to adopt American methods, they (the foreigners) became much more proficient at making standardized, interchangeable watch parts, and they eventually surpassed the Americans in producing high-quality watch tools.

In time, the Swiss could produce parts of equal quality and at a lower price.   The tariff on movements and parts kept the American producers in business.   The Underwood Tariff of 1912, set the tariff at 30 percent ad valorem.   In spite of this high rate, foreign producers could assemble a watch in the United States for less than could a domestic supplier.   The tariff was one reason, but not the only one.   In the 1920's, foreigners used much more advertising and they promoted sales with installment credit.   Thus, when the crash came in 1929, the Americans faced the depression in a relatively weak position. (49)

The Great Depression

The economic collapse that commenced in 1929 was complete. Prices and production fell nearly 50 percent, and this threatened to destroy many manufacturing industries, including watchmaking.   But things at Waltham did not worsen immediately. In fact, employment actually rose slightly in early 1930. By June 1930, however, the factory had closed and employment fell from 2,392 to 1,559. (50)

The Congress responded to the plight of American manufacturing by passing the Smoot-Hawley Tariff Act of 1930. President Hoover approved the legislation in spite of having received a petition from 1,028 economists urging him not to sign the bill. (51) The law increased the tariff on Swiss watches, but Swiss manufacturers still dominated the American market.

Because it was difficult to enforce the tariff law, smuggling of foreign movements was widespread. The Swiss were driven to smuggling because the Tariff exceeded the value of the movements being produced. It was estimated that Americans lost sales of up to two million watches per year because of smuggling. Dumaine tried to get Congressional support for limiting the importation of foreign movements, but while he was making the attempt, Franklin D. Roosevelt was elected President of the United States.

Roosevelt saw the problem quite differently. He believed, rightly, that foreign trade was a two-way street: that Americans had to import if they wished to export. Therefore, he ordered Secretary of State Cordell Hull to embark on a good-neighbor policy of reducing restrictions on foreign trade. The Swiss, sensing the change in United States attitude, petitioned the Tariff Commission for relief. The Commission responded favorably, granting an average reduction of 34 percent.

The National Industrial Recovery Act, passed on June 16, 1933, was the key to Roosevelt's recovery program. Its primary purpose was to eliminate unfair competition and overproduction, to restore profits to business and to provide a living wage. These objectives were to be achieved by the "Codes of Fair Competition." The code for the "Assembled Watch Industry" was approved on August 27, 1934. The code established minimum wages and maximum hours, but it also dealt with legally imported watches and movements and smuggling. (52)

When the tariff on Swiss watches and parts was reduced, Waltham had no choice but to reduce its prices. Faced with a declining market, competitive pressure, overproduction, and idle capacity, Dumaine fought for Waltham's share of the market. But lower prices were not enough, so the company returned to an innovative approach. It introduced two new baguette models--the 400 model and the 450, which were offered in 1931 and 1933, respectively. The 400 was advertised as the smallest "made-in-America" watch, but the 450 was more popular, because it sold for $18 to the consumer, compared with $30-100 for the 400. Sales of the 450 rose from 216 in 1933 to 20,341 in 1936. (53)

As we stated above, Waltham turned to making clocks in the mid-1920's, and in 1927, it produced its first electric clock. But the sales of electric clocks were insufficient to offset the decline in 8-day clocks and pocket watch sales continued the long-run decline that had started around World War I.

In 1936, Waltham earned a small profit, but the underlying condition of the company was not sound. As *Sales Management* stated, Dumaine "lifted Waltham out of the red and kept it operating profitably for 20 years," but "he cut wages, abolished jobs, cancelled advertising contracts, spent little on new machinery, research and new watch designs." He "skimmed through the depression on shutdowns." (54)

Waltham Produces for War A Third time

Before World War II, Waltham produced pocket watches, wrist watches, 8-day clocks, electric clocks, automobile clocks, tachometers, pickometers and precision parts. This production plan changed, however, when the company received an order in 1940 from the United States government for the production of "educational" time fuses. When war was declared, Waltham produced aircraft clocks, chronometers, compasses, stop watches, speedometers, rifle parts and other precision parts. The peak production was reached between 1942 and 1945. In this period, the company reached all the goals it set out to accomplish. It sold over $28 million of goods; its watches bettered the government standards for keeping time, and the company was put on a sound financial base. It paid off its bonds and reduced substantially its preferred stock. The company received an official stamp of approval when it received the Army-Navy "E" award--which stood for excellence in production.

The War Production Board, which controlled all production in World War II, civilian and military, ordered deep cuts in civilian output, but there was no curtailment of watch and clock production. It ruled that watches were semi-essential to the civilian economy. At the same time, Gruen, Bulova, and Longines-Wittnauer, which imported Swiss movements for assembly in the United States, continued to import throughout the war.

Waltham produced some wrist watches in the early months of the war--the 887 and 659 models, but after August 1942, the emphasis was mainly on timers and 8-day clocks (mostly CDIA). The CDIA (Civil Date Indicator Aircraft) were 8-day, 37 size, with 24 hour dials and sweep-second hands. They had black Bakolite cases. These clocks were introduced in the 1920's, but after 1936, numbers and letters were used to designate certain features (ie P for pendant set, 815 for 8-day and 15 jewels). Some aircraft clocks were model 22809--P or S, in 22 size, with 12 hour dials. They were unadjusted.

The timers were first manufactured in June 1943. They were

16 size, 9 jewels, open face, adjusted, and the sweep second hand made one revolution per minute. They could fly back to 12 with either the push pin or the crown. Most cases came in stainless steel. The 1/5 second timer shown in Figure 37 was produced in January 1944 for the British Admiralty.

FIGURE 37 *A Waltham Military Stop Watch*

Waltham produced timepieces for all the military services in World War II. For the Navy, it made deck watches, chronometers, and marine clocks. The chronometers were among the most accurate of watches, but they required special handling, because of their delicacy. Consequently, they were kept in a "secure place in the middle of the ship, where pitch and roll are at minimum." (55)

Since the chronometers were not moved, the Navy required also a mobile watch--called a deck watch. These watches were used to "transport" the correct time from the chronometer room to other parts of the ship, particularly to the bridge, where the Captain functioned. Many deck watches had a "hack" feature, which allowed for stopping and resetting. These deck watches were usually open face, 16 size, 21 jewels, with bold numerals and minute marks around the dials.

For the Army, pocket watches were issued in two grades, depending on the accuracy required. Grade I was high quality and was issued to the Transportation Corps and the Demolition Units. The Hamilton 992B was often used for these purposes. Most watches, however, were Grade II. They were 9, 11, 15 and 17 jewels. Waltham produced both Grade I and Grade II watches. The Grade II watch was 16 size, 17 jewels, with large arabic numerals and 60 minute marks shown on the rim.

The Air Force required more accurate timepieces. Each large aircraft (ie the Flying Fortress) had a navigational watch on board, which served the same purposes as the chronometer on ships. Although most of these watches were Hamilton 4992B models, the Waltham 23 jewel Vanguard also qualified for the U. S. Army Air Force. These watches were kept on the navigator's table and were secured in a round metallic case with springs that allowed the watch to take the effects of air turbulence without damage.

By 1944, the Waltham Watch Company seemed to be in a comfortable financial condition. But at the peak of this war prosperity, Frederic Dumaine cashed in, thereby jeopardising the the long-run viability of the company. He sold his interests to Union Securities Trust Company, New York, and Ira Guilden, a former Vice President of the Bulova Watch Company.

Guilden purchased Waltham on a grudge. He had been married to Bulova's sister, and when he divorced her, he was released by the company. He was determined, therefore, to show Bulova and the world that he could succeed at watchmaking against all the competition, especially Bulova.

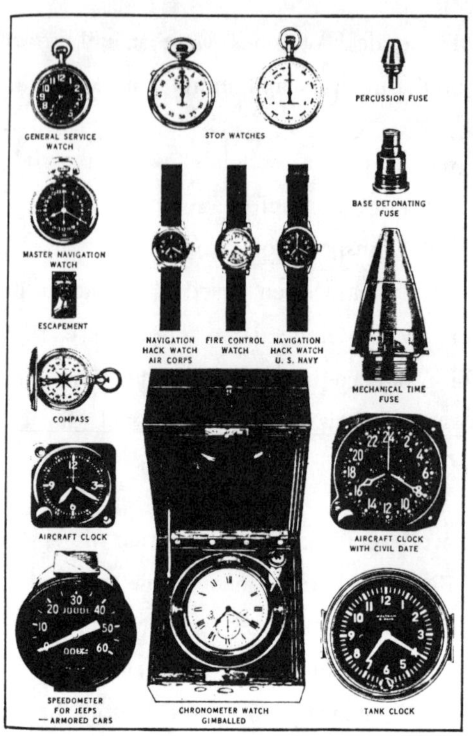

FIGURE 38   *Waltham Wartime Production, 1945*

Guilden was greeted in Waltham as a builder--a man of great energy and ambition. At 10 years of age, he had helped his widowed mother run a candy store in Brooklyn; at 17, he sold jewelry; at 26, he owned the Knickerbocker Watch Company; and at 32, he was made a Bulova Vice President. When he took over at Waltham, he was only 48 years old. (56)

One of his first moves was to inform the workers that they would have the finest machinery and the latest devices on which to work. He established a "Time Research" laboratory and instituted a new, much larger advertising program, under N. W. Ayer & Son. This included a new radio network series. In one year, advertising expenses rose more tham ten fold. At the same time, Guilden dropped all wholesalers and began to sell directly to "selected retail jewelers." This reduced his sources of distribution from 25,000 to 5,000.

Wall Street received this news favorably, but as financial analysts say, the fundamentals were not good. On December 28, 1948, the company filed for reorganization in the District Court of Boston. This action confirmed the rumors that Waltham was in serious financial trouble.

Who (or What) Killed Waltham

How did this dismal condition happen? (57) Did Dumaine kill Waltham? Why did he sell during a war prosperity? Was he leaving a sinking ship? Did he know something the public did not know? Or had he achieved all of his objectives?

In just four short years Waltham went from relative prosperity to bankruptcy. How this occurred was a matter of national concern. Should the Federal government allow for the failure of a company that had contributed mightily to the war effort and which employed one-fifth of all the workers in Waltham, Massachusett? (58)

Let us review the factors that eventually sank the company. First, while Waltham was completely immersed in war production, the Swiss achieved a near monopoly of the civilian market in the United States. But since a large share of these imports were requisitioned for the military, there was no great concern about the growing Swiss influence at the time. When World War II ended, the old tariff rates remained in place, but President Truman's "reciprocity" program allowed him to reduce rates up to 50 percent. Tariffs were, in fact, lowered soon after.

Second, American producers complained about wide variations in costs brought on by lower wages abroad. In 1945, American workers in the watch industry received two-and-one half times more than their Swiss competitors. Since wages were 70-90 percent of total costs, the discrepancy was a serious impediment for American companies. They sought a quota plan to reduce imports.

Third, there were a growing number of jewelers who complained about the quality of Waltham's watches. Despite claims by Guilden that "Waltham would make the finest American watch in every sense," movements from Waltham arrived in dirty or non-operating condition. For example, many of the 670 and 675 movements were not only dirty but the cases did not fit. For these reasons, the Waltham inventory of unsold watches mounted. (59)

The final factor affecting the long-run viability of the Waltham Watch Company was financial. The company was literally drained of its liquid assets. It started with Dumaine. When he sold out in 1944, he was not concerned with what would happen to Waltham when V-J Day arrived. Guilden made an honest effort to market Waltham watches, but his "enlarged and invigorated" sales force did not help. He persuaded banks to loan an additional $3.5 million, but then he too began quietly to sell his own shares.

When Guilden left Waltham, the Board of Directors chose Paul P. Johnson, of Thompson Brothers, to direct operations. He began a large advertising campaign to celebrate the "33,000,000th watch." (60) His advertising program was also to feature a Centennial Series, in anticipation of the 100th birthday of the company. This watch would have a mainspring that was "guaranteed to last until 2050 A. D."

Bailout Attempts

After Waltham petitioned the court for a reorganization in 1949, it appointed 3 trustees, and in April 1949, they were granted a $6 million RFC loan--one-third to be used for modernization. One of the trustees, John J. Hagerty, replaced Johnson as head of the company.

But Hagerty knew nothing about watches. He fired all the top executives of Waltham and revised the whole selling strategy. Waltham had sold 95 percent of its watches through its 5,000 jewelers (Dumaine cut the number from 25,000 to 5,000 earlier). Now Hagerty reduced it to nearly zero. He offered the entire inventory of watches to the Associated Merchandising Group (Filene's of Boston, J. L. Hudson's of Detroit, Bloomingdale's of New York City). The Company was able to sell its large inventory at one-half of its value in the department stores and to restore its cash position temporarily.

With some of the new cash, Hagerty planned an "All- American Watch" series. In a brochure, he previewed his watches for the next century: there would be a line for each holiday (a national defense series for Memorial Day, a salute to workers for Labor Day, etc.) and there would be a "Southern Belle" series for the ladies.

Hagerty reported in November 1949 that Waltham's sales were reaching expectations, but on February 3, 1950, the company was shut down again. At this point, the RFC took over, since it was the leading creditor.

For the next two years, Waltham, this time joined by Elgin and Hamilton, petitioned the White House to limit Swiss imports. On August 14, 1952, President Truman turned down their request. This action called for a new strategy. Waltham announced it would import 17-jewel movements from the Swiss, and it would concentrate on manufacturing 19- and 21-jewel movements. (61) This was a curious deviation from a long-standing trend; in most manufacturing, the United States provided the low price mass-produced items and foreigners produced the high-cost, quality items.

In 1953, a new issue was injected by the United States government: the Justice Department claimed that the Federation of Watch Manufacturers in Switzerland violated the Sherman Antitrust Act by controlling production. marketing and the selling prices of Swiss watches. (62) This offered little hope to Waltham. however. Between 1947 and 1957. it had 6 different managements; they did not generate enough profits to allow the managers many options. Finally. in 1957. a new group took control of Waltham from the Bellanca Corporation. The company was split in two. The Waltham Watch Company became the Waltham Precision Instrument Company, and a new company, Waltham Watch Company of Delaware, was set up within the Instrument Company to be spun off later.

FIGURE 39   *One of the last Walthams--a battery-operated model*

The new Waltham Company made gyroscopes. timing devices, and aircraft clocks. But its watch operation was continued only to utilize its large tax-loss credit. What an ignominious end for the company that gave us the genius of Aaron Dennison. Charles Moseley. Charles Vander Woerd and hundreds of other great craftsmen!

References

1. The most complete treatment of the Waltham Watch Company is Charles W. Moore, *Timing a Century--History of the Waltham Watch Company* (Cambridge: Harvard University Press, 1945), but a good summary may be found in Edmund L. Sanderson, *Waltham Industries* (Waltham: Waltham Historical Society, 1957). Moore's approach is a business history, which was done for the Harvard Business School. Its main objective was to outline the management policies that helped or hindered the company in achieving a dominant place in the industry. Our approach is to deal with the company as a producer of watches, and it is aimed mainly at collectors.

2. Moore, *Timing a Century*, p.20.

3. Wesley R. Hauptman, "Appleton, Tracy and Company," NAWCC *Bulletin*, X (Apr. 1963), 690.

4. Moore, *Timing a Century*, p.29.

5. Hauptman, "Appleton, Tracy," p.694.

6. The numbers here, too, are conflicting. Moore states that Robbins "sold the watch factory to the Land Company for $125,000." (p.36). Sanderson mentions no price, but Abbott places the entire capital at $300,000, with the watch company putting in $225,000 and the Improvement Company $75,000.

7. Moore, *Timing a Century*, p.38.

8. Hauptman, "Appleton, Tracy", p.94.

9. See Thomas L. De Fazio, "The Nashua Adventure and the American Watch Company," NAWCC *Bulletin*, XVII (Dec. 1975), p.575.

10. Henry G. Abbott, *History of American Waltham Watch Company* (Chicago, 1905), p.23.

11. Frederick Mudge Selchow, "The Nashua Watch Company," NAWCC *Bulletin*, XVII (Dec. 1975), p.549.

12. De Fazio, "The Nashua Adventure", p.585.

13. George E. Townsend, *Almost Everything You Wanted to Know About American Watches and Didn't Know Who to Ask* (1971), pp.40-42.14.

14. Roy Ehrhardt, *Waltham Pocket Watch Identification and Price Guide* (1976), p.8.

15. Wesley R. Hauptman, The American Watch Company," NAWCC *Bulletin*, XI (April 1964), pp. 191-192.

16. See August C. Bolino, "The Elusive Model A," NAWCC *Bulletin*, XXIV (April 1982), 167-168. The comments of Willard W. Halsted and Michael C. Harrold were expressed to me in letters following the publication of this brief article.

17. Theodore C. Cuss, *Cammerer Cuss Book of Antique Watches* (Baron, 1976), p.101.

18. Moore, *Timing a Century*, p.40.

19. Abbott, *American Waltham Watch Company*, p.31.

20. Orra L. Stone, *History of Massachusetts Industries* (Boston, 1930), p.946.

21. "Eighty Years' Progress of the United States: A Family Record of American Industry, Energy and Enterprise," 1868, Reprint, NAWCC *Bulletin*, August 1981, p.373.

22. I wish to thank Taff for some of the information in this section, which he provided in a letter dated December 15, 1986.

23. Larry Treiman, "Answer Box," NAWCC *Bulletin*, XVIII (June 1986), p. 255.

24. Moore, *Timing a Century*, p.52.

25. Hauptman, "Appleton, Tracy," p.186.

26. Moore, *Timing a Century*, p.55.

27. *Fitchburg Daily Sentinel*, July 17, 1875.

28. J. E. Coleman, "Clocks, Watches Collectors, People, and Legends," NAWCC *Bulletin*, XI (October 1964), p.430.

29. Moore, *Timing a Century*, p.65.

30. Moore, *Timing a Century*

31. Warren H. Niebling, "A History of American Watchcase," NAWCC *Bulletin*, 1969-1970, pp. 627-633.

32. Joseph E. Brown, "The Invention of an Improved Compensation Balance: The Mechanical Genius of Charles Vander Woerd," NAWCC *Bulletin*, XXIV (October 1982), 487-493.

33. Quoted in Hauptman , "Appleton, Tracy," p.201.

34. American Waltham Watch Company, *The Perfected American Watch* (Waltham, 1904), p. 31.

35. Moore, *Timing a Century*, p.77.

36. Michael C. Harrold, *American Watchmaking: A Technical History of the American Watch Industry, 1850-1930*, Supplement, NAWCC *Bulletin*, Spring 1984, p.44.

37. Robert N. Conley, "Waltham: A City Becomes a Pocket Watch," NAWCC *Bulletin*, XXIII (Oct. 1981), 482.

38. "Watch Trade Trust," *The New York Times*, July 26, 1903, p. 11.

39. In 1963, after the Waltham Watch Company was sold, a second suit was filed, this time by the Federal Trade Commission, over use of the Waltham name. In this case, European clocks were being imported and the Waltham name was attached to them. The FTC stated that this practice was deceptive, because the sellers were giving the impression that these clocks were made by the Waltham Watch Company.

40. Peyton Autry, "Waltham's Magical Maximus," NAWCC [Bulletin,] XXII (Oct. 1980), 531-539.

41. Steve Lindberg, "The Production History of the Maximus,"

NAWCC *Bulletin,* XXVII (Apr. 1985), Table 1.

42. Moore, *Timing a Century,* p.106.

43. August A. Berle and Gardner Means, *The Modern Corporation and Private Property* (New York: Macmillan, 1933).

44. Vincent P. Carosso, "The Waltham Watch Company," *Bulletin of the Business Historical Society,* XXIII (Dec. 1949), p. 175.

45. Moore, *Timing a Century,* p. 134.

46. The history of wrist watch production is treated in Chapter 12.

47. Vernon Hawkins, *Movement Production: American Waltham Watch Company* (Kansas City: Heart of America, 1982).

48. Roy Ehrhardt, *American Pocket Watches, Encyclopedia and Price Guide* (Kansas City: Heart of America, 1982) I, p. 65.

49. Moore, *Timing a Century,* p. 214.

50. Moore, *Timing a Century,* p.207.

51. August C. Bolino, *Development of the American Economy,* (Columbus: Charles E. Merrill, 1966), p. 476.

52. See U. S. National Recovery Administration, *Code of Fair Competition for the Assembled Watch Industry* (Washington, 1934).

53. Moore, *Timing a Century,* p. 217.

54. Lawrence M. Hughes, "Who Killed Waltham?" [Sales Management,] April 15, 1950, p.118.

55. R. C. Williamson, "Military Timekeepers," NAWCC *Bulletin,* XIV (Feb. 1973), p. 252.

56. *Sales Management,* April 15, 1950, p. 120.

57. August C. Bolino, "Who (Or What) Killed Waltham?" NAWCC *Bulletin,* XXV (October 1983), 555-557.

58. Carosso, "The Waltham Watch Company," p. 166.

59. *Business Week*, September 8, 1945, p. 54.

60. Hughes, "Who Killed Waltham," p. 120.

61. *Wall Street Journal*, March 20, 1952.

62. *Business Week*, December 26, 1953, p. 70.

CHAPTER 10:
THE WALTHAM WATCH COMPANY--
CONTINUED

In this chapter, we complete the Waltham story by discussing some interesting Waltham watches, Waltham movement production and the Waltham contribution to horological history.

Some Interesting Waltham Watches

Because the Waltham Watch Company made approximately 35 million watches, it is difficult to single out particular watches. But some watches are special, and we describe them here briefly.

As Table 12 shows, the Waltham railroad quality, up and down, indicators were produced from 1913 to 1928. Most were 16 size, 23 jewels, 1908 model Vanguards. But some were 1899 and 1912 models, and some had 21 jewels. Robert Porter was surely understating when he called these watches an "interesting mechanism." (1) He calculated that a full turn of the rachet wheel stored up 6.5 hour of running time in the mainspring barrel, and that, on the downside (the unwinding), a 360 degree rotation moves the hand just over 61 degrees. Thus, he computed that Waltham required 144 degrees of arc for the indicating dial.

The watch shown in Figure 40 is a 16 size, 21 jewel, 1908 model, Crescent Street, serial number 20,142,564 (manufactured in 1914). It has a 5-60 minute arabic dial and a Keystone, J. Boss, 10 carat gold filled case (9,391,042). It is adjusted to 5 positions.

A second interesting Waltham watch is the chronograph, because it is so scarce. The first chronograph was introduced in 1877. It was a standard 14 size, with chronograph attachments, that were patented by H. A. Lugrin, of New York, on June 13 and October 23, 1876. He supervised the manufacturing of these parts in the Waltham Building on Bond Street, New York. But just as production

– 172 –

Table 12

WALTHAM UP AND DOWN INDICATORS

Serial Number*	Model	Year Produced	Type	No. of Jewels
			16 Size	
19,403,001	1899	1913	Maximus	23
18,010,001	1908	1910	Vanguard	23
18,074,501	1908	1910	Vanguard	23
18,131,501	1908	1911	Vanguard	23
19,104,501	1908	1913	Vanguard	23
20,012,801	1908	1915	Vanguard	23
20,091,001	1912	1915	Crescent Street	21
20,102,001	1908	1915	Crescent Street	21
20,155,501	1908	1915	Crescent Street	21
22,001,001	1908	1918	Crescent Street	21
22,020,001	1908	1918	Vanguard	23
22,053,501	1908	1918	Vanguard	23
22,064,501	1908	1918	Vanguard	23
23,100,001	1908	1919	Maximus	23
23,152,001	1908	1919	Vanguard	23
23,170,001	1908	1919	Vanguard	23
24,403,001	1908	1923	Vanguard	23
24,568,001	1908	1924	Vanguard	23
24,604,001	1908	1924	Vanguard	23
25,255,001	1908	1926	Vanguard	23
25,534,001	1908	1926	Vanguard	23
25,401,001	1908	1926	Vanguard	23
25,511,001	1908	1926	Vanguard	23
26,560,001	1908	1928	Vanguard	23
26,615,001	1908	1928	Vanguard	23
26,665,001	1908	1928	Vanguard	23
26,745,001	1908	1928	Vanguard	23
			8 – Day	
19,175,001				7
19,447,001				7
19,569,001				7
19,975,001				7
20,093,001			Chronometer	15
20,231,001				7
20,298,001				7
23,293,001				7

*Denotes beginning of production run only

Source: Roy Ehrhardt, <u>Waltham Pocket Watch Identification and Price</u> (1976).

FIGURE 40   *Waltham UP and Down Indicator*

was to commence, in March 1877, the building caught fire, which delayed production until the end of the year.   (2)

There were several types of chronographs, but usually they were either the donut ring on the front dial type, with sweep-second fixed to the back, or they had regular dials and hands.   The earliest of Waltham's chronographs were of the donut ring-type.

Because of the manner in which the chronograph was made, there is some confusion about dates.   The attachment was added to many movements, after it was patented.   Table 13 shows that chronographs were produced from 1877 to 1888.   (3) Most were 14 size, three-quarter plate, and had 13 jewels (although nearly as many 15 jewel movements were made.)   The original 1874 model was replaced in 1884, but Ehrhardt lists model runs of 1873 and 1877 in 1878 and 1883.   The last of the chronographs was made in 1888 as a 16 size, 17 jeweel model.

The Waltham repeater is the third of our watches to be

Table 13

WALTHAM CHRONOGRAPH PRODUCTION

Serial Number*	Size	Model	Plate	Jewels	Year Produced
1,015,901	14	1874	3/4	16	1877
1,050,951	14	1874	3/4	16	1877
1,090,001	14	1874	3/4	7-13	1877
1,227,701	8	1873	3/4	7-11	1878
1,609,801	14	1874	3/4	15	1880
1,621,301	14	1874	3/4	15	1881
1,833,001	14	1874	3/4	15	1881
2,073,001	14	1874	3/4	13	1883
2,073,201	14	1874	3/4	13	1883
2,343,201	14	1874	3/4	13	1883
2,343,951	14	1877	3/4	15	1883
2,360,901	14	1874	3/4	15	1884
2,379,901	14	1884	3/4	15	1884
2,403,001	14	1884	3/4	13	1884
2,469,001	14	1884	3/4	13	1884
2,718,001	14	1884	3/4	13	1885
2,719,801	14	1884	3/4	13-15	1885
2,720,001	14	1884	3/4	15	1885
2,809,901	14	1884	3/4	15	1885
2,823,001	14	1884	3/4	13	1885
3,037,001	14	1884	3/4	15	1885
3,037,101	14	1884	3/4	13-15	1885
3,037,801	14	1884	3/4	13	1885
3,092,601	14	1884	3/4	13	1885
3,126,001	14	1884	3/4	15	1886
3,126,501	14	1884	3/4	13	1886
3,159,301	14	1884	3/4	13	1886
3,160,001	14	1884	3/4	13	1886
3,793,001	16	--	--	17	1888

*Only beginning number of production run shown

Source: Roy Ehrhardt, <u>Waltham Pocket Watch Identification and Price</u>, 1976.

highlighted here. On August 21, 1882, Fred Terstegen applied for a patent for a repeating attachment for watches. He wanted to produce an attachment that could be applied to any American movement, without altering the shape or size of the watch. He was granted three patents (January 27, 1885, February 18, 1890 and September 9, 1890) that he claimed would allow for one-hour, quarter-hour, five-minute or one-minute repeaters. He founded the American Repeating Watch Company at 117 Broad Street, Elizabeth, New Jersey. It is not known how many attachments were actually produced. Nijssen tells us that, "The Waltham Watch Company is the only company on record who has fabricated a repeating watch in America, but it was of Swiss design and the parts were made in Switzerland." He was able to identify six different production runs of repeaters. (see Table 14)

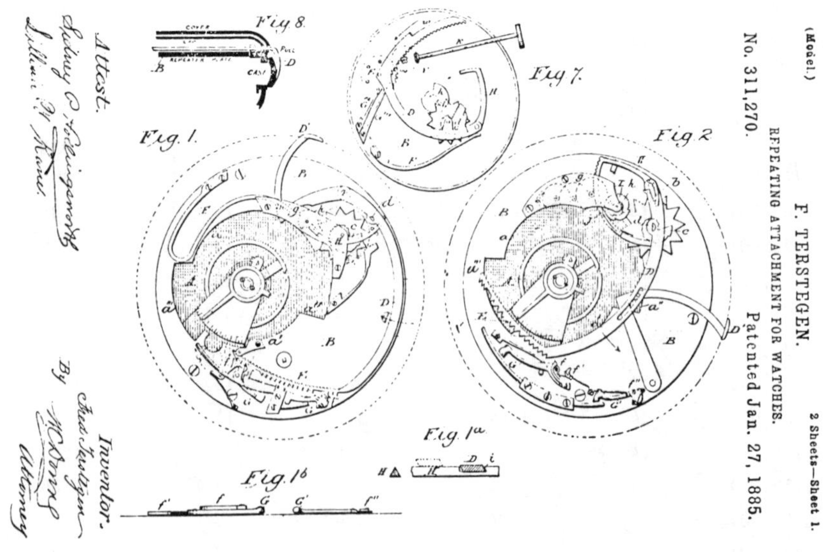

FIGURE 41 *The Terstegen Repeating Attachment*

He adds that, "It will be impossible to find out how many were ever made." (4) Later, De Fazio attempted the impossible. His

Table 14

PRODUCTION OF WALTHAM REPEATERS

	Serial Number		Model	Grade	Style	Qty
1)	#2605701	2606000	1872	Riverside	Htg	299
2)	2809501	2809700	1884	Am. W. Co.	Htg	199
3)	3037101	3037700	1884	Chron	Htg	599
4)	3127101	3127700	1884	Chron	Htg	599
5)	3793001	3793400	--	Chron	Htg	399
6)	3793401	3794200	--	Repeater	Htg	799

Source: Gerrit A. Nijssen, "The American Repeater," NAWCC _Bulletin_, XVI
(December, 1973), p. 15.

survey of repeaters leads him to state that "perhaps as many as 1,250 to 1,350 of these watches, using the Charles H. Meylan repeater and/or chronograph patents, appear to have been finished." (5)

De Fazio's figure is obviously a "guesstimate," because regular watches, chronographs and repeaters were mixed in the same lots. His survey tells us that repeaters came in 14 and 16 sizes and that some were chronographs. The chronograph/repeaters were particularly complicated, using the Lugrin patent. These were essentially Swiss watches. The "American" counterparts were provided by Charles H. Meylan. His repeaters were finer and had quality finishes, including damascening of the balance cock, balance wheel and fourth wheel. The quotes in the word American (above) are there because the Meylan family produced watches in the Lake Joux region of Switzerland "since the beginning of watchmaking in the valley." (6) But Charles H. Meylan became one of a famous group of Swiss who produced watches in the United States before the turn of the twentieth century.

The next of our special watches is presented in Figure 42. It is a 16 size, 20 jewel, 1899 bridge model that was made for Smith, Patterson Company, a leading Boston jeweler. The serial number 14,050,891 indicates that it was made in 1905. The watch is shown in its original mahogany box.

The last watch in this section is a curiosity. It is a "Wm Ellery," number 5364. It looks like the usual 1857 model, with its solid balance wheel and straight balance cock. But the William Ellery is a Civil War watch, the first production run being 46,201 (1861). Is 5364 a fake? I disassembled it and some Waltham parts fit. The only oddity is that the regulator is usually to the left of the balance cock on the Wm Ellery model, while on 5364, it is to the right. If it is a fake, it is a darn good one. You can see for yourself. The left part of Figure 43 is number 5364; the right side is William

FIGURE 42 *The 16 size, 20 jewel, 1899 bridge model Waltham*

Ellery number 394200 (1868). It is 15 jewel, adjusted, and has a Fahy's Monarch Coin Silver case (S 8374).

FIGURE 43 *Two William Ellerys Compared*

Movement Production

Recently, Vernon Hawkins transcribed all the annual reports of the Waltham Watch Company for the years 1859 to 1899. In reviewing the production statistics found in these annual reports for the nineteenth century, we have made a discovery: that the production of Waltham watches is overstated in the horological literature. Ehrhardt shows a total of 9 million for 1899. (7) If you add Townshend's figures (which we did) the sum is 9,558,361. (8) Shugart presents a lower total for the century: 8,370,000. (9) The production statistics shown in Table 15 are those presented by Royal Robbins. Summing his figures we obtained a grand total of 8,053,770 less than all publications offer to date. As we can see, Shugart's figure is closest to the calculated sum. The obvious conclusion is that these movements are scarcer than we were led to believe. But some models are much scarcer than others.

In Table 16, we offer Vernon Hawkins' statistics for the production of Waltham movements, shown in descending order. We can see easily which models are scarce and not scarce. As we all know, the 1883, 18 size model is the most abundant. On the other end, we see the scarcest Walthams: the Special 10 size (key wind), the chrono-repeater, the 1891 0/0 size, the 1875 12 size, the 1868 16 size, the 1862 20 size and the 1900 10 size--all with total production of less than 5,000 movements. It is also clear that the "age" of the movement has no relationship to its scarcity. We are surprised to note that there were 836,067 movements of the 1857 model and only 32,550 of the 1904 14 size. Of course, as Hawkins shows, within each of the aggregates, there are some very scarce model runs. One obvious example is the J. Watson grade of the 1857 model, of which only 1,200 were produced. Another would be the Canadian Pacific grade of the 1883 model, with a total production of 1,020.

A few comments on the data. First, since the data in Table 15

Table 15

WALTHAM MOVEMENT PRODUCTION, 1859-1899
(Feb. 1 - Jan. 31)

Year	Number of Movements Produced	Year	Number of Movements Produced
1859	12,304		
1860	12,055	1880	176,117
1861	2,734	1881	220,590
1862	19,059	1882	255,194
1863	38,103	1883	298,240
1864	44,632	1884	311,070
1865	72,831	1885	270,857
1866	64,482	1886	301,843
1867	42,089	1887	340,733
1868	52,168	1888	372,329
1869	55,042	1889	413,290
1859-1869	415,499	1880-1889	2,960,263
1870	66,655	1890	459,631
1871	74,530	1891	510,733
1872	79,346	1892	512,735
1873	64,847	1893	422,855
1874	49,243	1894	325,609
1875	84,737	1895	389,447
1876	82,053	1896	339,033
1877	96,447	1897	364,421
1878	98,776	1898	515,518
1879	141,392		
1870-1879	838,026	1860-1898	3,839,982

Grand Total 1859-1899   8,053,770

Source: Vernon M. Hawkins, <u>Waltham Watch Company, Annual Reports to Stockholders, 1859-1899</u> (West Boxford MA, 1984)

Table 16

WALTHAM MOVEMENT PRODUCTION
(Decreasing Order)

Model	Size	Number Produced
1883	18	5,417,467
1894	12	3,256,542
1908	16	2,640,376
8-day	36	1,801,700
1890	6	1,632,209
1899	16	1,582,274
1891	0	1,520,837
1898 1912	6/0	1,263,243
1877	18	1,087,333
1857	18	836,067
1888	16	806,918
1907	0	727,290
1900	0	564,563
Colonial	14 12	558,985
1884	14	480,376
1897	14	340,751
1892	18	327,471
1873	6	282,212
1873	8	270,520
1870	14	232,400
Equity	16	113,300
1889	6	108,700

1879	18	108,375
1874	14	106,349
1895	14	106,100
1882	1	96,000
Ball	16	85,200
1882	0	79,300
1872	16	50,685
1874	10	37,547
1865	10	34,475
1870	14	33,690
1904	14	32,550
1859	18	31,940
1870	18	17,500
1873	1	14,000
1861	10	11,600
1860	16	11,451
1878	10	8,670
1900	10	4,000
1862	20	3,549
1868	16	2,500
1891	0/0	1,800
Chrono-repeater	16	1,200
Special	3	500
Special	10 (KW)	400
1875	12	200

Source: Vernon Hawkins, <u>Movement Production</u>, American Waltham Watch Co. (Kansas City: Heart of American 1982).

are from the annual reports, we must assume that they are correct. However, the method of dating seems to change in the reports. For example, the 1860 figure is for the 12 months from January 31, 1859 to February 1, 1860, but the 1861 total is "During the year 1860."

Second, Robbins gives no figure for 1861, but one can be computed by using production and inventory data. The end of year inventory for 1859, 1860 and 1861 total 4,765, but the inventory on hand given by Robbins is 4,654--a discrepancy of 111 movements.

Third, no production statistics are presented in the 1893 report. Page 71 of Hawkins gives the 1891 and 1892 output and page 75 gives the 1894 and 1895. But since Robbins offers his total production for the years 1859 to 1894, we can obtain the movement production for 1893 by subtraction. We obtained the 1869 estimate in the same way, since Robbins showed total production for 1858 to 1870 as 415,499.

The Waltham Contribution

How shall we rank Waltham in the annals of horology? The answer is given, in part, in Chapter 9. Nearly all watch companies that arose to compete with Waltham were formed with Waltham alumni. These men, Stratton, Bartlett, Woerd, Bigelow, Gerry and Moseley make up a Who's Who of watch making. The list of innovations emanating from Waltham is long and impressive.

The designers, inventors and mechanics who labored for the Waltham Watch Company established principles of manufacturing that came to be used by other industries, including heavy industry. Some of their improvements include the application of spring clutches to lathes, the automatic indexing mechanism for cutting pinions, the design of measuring instruments for jewels and settings, the creation of screw-cutting machines and the development of mechanisms to transfer work from one station to another. These innovations would

have been impossible without the long list of skilled machinists. many
provided at the Waltham plants.

References

1. Robert D. Porter, "Waltham's Up and Down Work," NAWCC *Bulletin*, XXVII (October 1985), p. 577.

2. Hauptman, "American Watch Company," p. 202.

3. The Ehrhardt list, shown in Table 12, is obviously not complete, because Hauptman offers a photo of a 14 size chronograph movement (number 987,009) not listed in Ehrhardt. This would put the beginning at 1876 instead of 1877.

4. Gerrit A. Nijssen, "The American Repeater," NAWCC *Bulletin*, XVI (December 1973), p. 5.

5. Thomas L. De Fazio, "The American Waltham Watch Company Repeaters," NAWCC *Bulletin*, XVIII (June 1977). p. 271.

6. De Fazio, "Repeaters," p. 277.

7. Ehrhardt. *Encyclopedia and Price Guide*, (1982), p. 19.

8. Townsend, *Almost Everything You Wanted to Know*, p. 40.

9. Cooksey Shugart, *The Complete Guide to American Pocket Watches* (Cleveland: Overstreet, 1981), 95.

PART III

Some Special Topics

In early railroad history, time was not an important consideration. Often trains ran on the same track in opposite directions. The first train to arrive waited for the next one to pass. But by 1860, when the watch industry in the United States was struggling to survive, there was already a well-developed network of railroads. Total track miles were 30,626, and between 1860 and 1893, five transcontinental lines were constructed to the west coast. (1) In a very real sense, the watch industry owes its growth to the spread of railroad service, and particularly to the increase of speed.

Establishing Standard Railway Time

In 1872, at a meeting of railroad superintendents in St. Louis, Missouri, the first draft of a "Time Table Convention" was drawn. Finally, on November 18, 1883, Standard Railway Time was accepted voluntarily by most United States and Canadian railroads. Before this, there was a confused variety of local times, which often differed by minutes or seconds from each other (because they were derived from solar positions). Because of its inherent advantages, most localities adopted the Railway Time, but it did not become federal law until March 1918, when the United States Congress was interested in conserving fuel during World War I. (2)

But the use of Standard Railway Time depended on the development of certain innovations so that time in one place could be coordinated with time elsewhere. These inventions perfected a system of electro-mechanical devices that allowed for the transmission of clock beats telegraphically over long distances. Thus, Standard Railway Time is linked up to the development of the telegraph. It made it possible to know the time at the same instant in New York and San Francisco.

The story of this achievement begins on June 9, 1844, 17 days after the start of the first telegraph. The United States Navy, on that date, telegraphed the correct time between Washington D. C. and Baltimore. It is particularly noteworthy that Harvard University set up a system from 1839 to 1859 to distribute the correct time from its observatory clocks to the entire Boston area, so that the Boston Watch Company could make use of this correct time in adjusting its watches.

By 1870, there were several observatorics dispensing the correct time; in fact, Samuel P. Langley inaugurated a "Time for Sale" program in that year. By then, manufacturers were keenly interested in the process, and E. Howard & Co. studied the potential of refining the time-keeping process.

Between 1850 and 1880, several patents were registered for keeping accurate time using electric clocks. By 1890, electrical impulses were used commonly for timekeeping purposes, and several private companies began to supply time-distribution systems, including Standard Electric Time Company of New Haven and the Time Telegraph Company of New York.

The Railroad Watch

The first improvement in railroad time keeping was the establishment of standard time, which was agreed upon by most railroads in 1883. The nation was divided up into 4 time zones and the public adhered to this method of timekeeping, although it was not made official until later.

The next improvement, and what concerns us here, was the design and manufacture of more accurate timepieces. In fact, the watch companies responded early to the demand for more accuracy. In 1866, the American Watch Company contracted with the Pennsylvania Railroad Company to supply it with about 300 18 size, 1857 model, 15 jewel Appleton, Tracy & Co. watches, which were marked "PENNA RAILROAD" on the dials. These were cased in silver with the number of the locomotive on the dust cover. In the early 1870's, the Pennsylvania Railroad made a similar contract with the Elgin Watch Company. According to Meggers, 600 of the B. W. Raymond models were purchased by the railroad, and as he stated, "These Waltham and Elgin made customised watches are the only ones that were purchased directly by a railroad." (3)

The Hampden Watch Company was the second Massachusetts company to supply railroad watches, these being the 18 size, 15 jewel, key wind "Rail Way," which came on the market in 1877. A Railway model was produced by Hampden in its first year of operation and until its last year. But these early watches were just the beginning. The real growth in the production of high grade watches began with watch inspection.

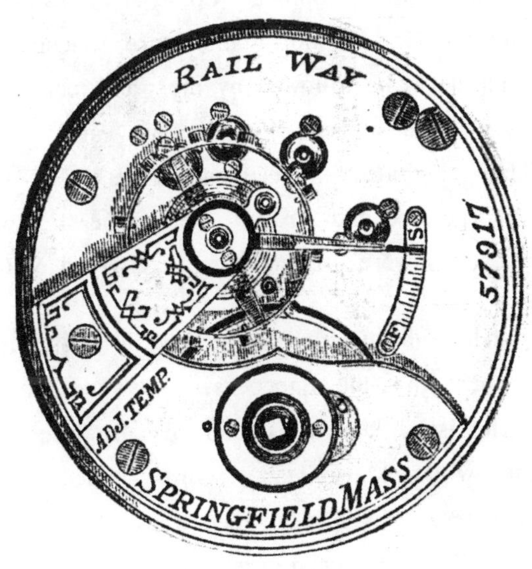

FIGURE 44 *The Hampden Watch Company "Rail Way" model, 1877*

We all know the story of the wreck of the two trains from the Lake Shore and the Michigan Southern Railroads (about 30 miles east of Cleveland, Ohio) at Kipton, Ohio on April 19, 1891. Although there was some watch inspection before the crash, it certainly hastened the spread of the practice. For example, J. C. Adams conducted a system of inspections for the Atchison, Topeka & Santa Fe Railroad about 1887. (4)

After the Kipton disaster, Webb C. Ball, of Cleveland, Ohio, was commissioned by the railroads to establish a uniform "time inspection" system. Eventually, he became the General Inspector for more than 125,000 miles of railroad in the United States. Canada and Mexico. (5)

When Ball surveyed the railroad timekeepers of northern Ohio, he found that engineers and conductors were using any kind of watches, even some they had received free for other purchases. And some were using alarm clocks! (6) Ball reacted to this survey by forming his own company--Ball & Co. (later changed to the Ball Watch Company). He did not manufacture watches, but rather he established standards to be followed by the watch companies. Ball had an excellent team that assisted him in this early venture into time service. They included Superintendent L. N. Cobb, Charles P. Gerdum, George Lee and Arthur Abel. The first watches to meet his new specifications were produced by E. Howard & Co.

Every watch with the Ball name had to pass inspection in his shop. It was here that the Babcocks, Mary and Harrison, played a major role. Harrison, known as Bab, usually tore down the watch and rebuilt it, while Mary kept records, tested, wound, set and adjusted the watches. She had to keep track of several watches, each of which was in the shop for a week or more. Her tests were very demanding. she put the watches on a tray and placed them in an oven set at 90 degrees; then she removed them and placed them in an icebox that was controlled at 40 degrees. Then the watches were ready for the five position test. If the watch passed all tests, it was cased, put in a chamois pouch and shipped to time inspectors all over the world. At its peak, the Ball organization was inspecting the watches of 108 railroads--approximately 800 watches. (7)

What is a Railroad Watch?

The chief question facing Webb Ball in setting standards for railroad use was, what watches would qualify as a railroad watch? When Ball made his determination, the watch companies had to produce these timekeepers (which they did in sufficient numbers). In fact, the Hamilton Watch Company was created in 1892 specifically to meet this new demand for precision watches, and between 1893 and 1969, when the last Hamilton railroad watch was made, they were the most desired timepieces of railroad engineers and conductors.

In discussing these standards, we start with the open face case. In the early days, some hunting cases were accepted, but these were rejected in favor of open face with a railroad dial. The two most common were the Montgomery, with each minute numbered on the outer rim, and the Ferguson, which had five-minute markers in black and the other numerals in smaller red figures. Some dials (ie Canadian) had the 13-24 hour designation, often in red. But the primary requirement was legibility. The numbers had to be bold arabic for easier reading.

Railroad watches were always produced with micrometric regulators. Waltham used the Church star-wheel regulator (patented in 1892) until the 1908 Patent regulator became available. Other companies used the Reed Patent regulator. These were needed for resetting watches when they drifted from the established 30-second per week requirement.

There was a confusing array of jewels in railroad watches. A fully jeweled watch had 17 jewels, but the most common railroad watches had 21 jewels. There was no argument about the basic 17 jewels. These included hole and cap jewels on the balance pivots, a roller jewel, pallet stones, pallet arbor jewels and jewels on the escape, center, third and fourth wheels. The confusion came when additional jewels were installed. Some were placed on the going barrel parts; others were added to the escape wheel or the pallet arbor. When this was done, the watch had 21 jewels. A Waltham representative suggested the futility of adding jewels when he stated that the 19 jewel Riverside grade "was the best- running timepiece." (8) The high-jeweled watches (23-26 jewels) were probably a sales feature, not a technical improvement. The railroad standards were further confused by the fact that although a 17 jewel wrist watch could qualify, often a 17 jewel pocket watch could not (except for those early ones which did).

All Ball Official Railroad Standard watches had double rollers and were adjusted to at least 5 positions (pendant up, right and left and dial up and down)

In summary, then, a railroad watch usually was open faced, in 16 or 18 size, with a minimum of 17 jewels. It was adjusted to 5 positions, had double rollers and a micrometric regulator that controlled the error in the watch to less than 30 seconds per week.

Watch Inspection and Time Service

Time Service refers to the procedures and regulations pertaining to watch inspection. Webb Ball established these time inspection services on most railroads by 1902. But watch inspection certainly existed before this. The Erie Railroad Company, for example, issued an order, effective February 12, 1888, that required conductors and engineers to "carry reliable watches." Under the order, certificates of inspection were mandatory, and they were to be renewed every six months. When the first round of watches was inspected, a large percentage was condemned, including some $300 gold chronometers. Since the employees had to purchase and maintain their own timepieces, this made the order "a pretty costly measure." An official of one line noted that as a result of the order the railroad would need about 2,000 quality watches at a total cost of $60.000 to $70,000. (9)

Given this demand for accurate timepieces, Ball was happy to service these lines, because he employed tradesmen inspectors in towns which were division points or railroad junctions. They inspected, adjusted and repaired railroad watches. Later they came to be authorized dealers in "Ball & Co." watches. In the next 20 years, until his death in 1922, Ball was able to provide time services for the Vanderbilt railroads (New York City) to ensure reliable, high-speed passenger service east of Chicago.

In 1922, the Southern Pacific Railroad established its own time service bureau, which was responsible for the supervision of 250 local watch inspectors, who maintained railroad watches for its employees. Its repair shop maintained 7,000 pendulum-type clocks for the Southern Pacific Railroad system. The rules were outlined in its "Time Service Manual." This manual made the cleaning of watches mandatory every years for engineers and every 4 years for switchmen. It required that watches keep time within 5 seconds per day, and that they be reset when they were off by 20 seconds. The railroad obtained the correct standard time daily by Western Union from the Naval Observatory in Washington, D. C. By contrast, the Union Pacific Railroad obtained its standard time from radio station WWV, the short-wave station of the National Bureau of Standards.

The National Bureau of Standards tests various types of measuring instruments, including watches and other time pieces (pocket watches, clocks, stop watches, chronometers and chronographs). The methods and procedures for testing timepieces is described in detail in a Bureau Circular. (10) The tests furnish information about the performance of various grades of watches. It is important to note that the Bureau did not test wrist watches, because "these watches are subjected to shocks and sudden changes of position which may affect their performance and cause the rate to change from day to day." The great variation in sizes was also noted as a problem in setting standards. (11)

The test of watches was to determine the daily rate of operation under various conditions for which they are adjusted. These include the position test, the temperature test and the isochronism test. The position test was to run the watch for several days in each of 3-5 positions. The temperature test included running the timepiece at 5, 20, and 35 degrees centigrade (41, 68, and 95 degrees fahrenheit). Isochronism tests whether the watch runs down at uneven rates. The adjustment is aimed at making the watch run evenly for the first 24 hours. All of the above standards were more rigid for railroad watches, stop watches and chronometers.

Massachusetts Railroad Watches

In a previous section, we mentioned the earliest railroad watches made in Massachusetts--the American Watch Company model of 1866 and the Hampden Rail Way model of 1877. These were the pioneers, but most railroad watches were made after 1893-- when the official standards became public.

E. Howard & Co. reacted to the growing demand for precision watches by making Ball watches. They were 18 size, 17 jewels and they had hunting cases. These were produced until 1895, and as Townsend stated, "I can only confirm 250 watches being made for Ball & Co." (12)

When the new E. Howard Watch Company was purchased by the Keystone Watch Case Company, of Philadelphia, it purchased the Howard name only and began to produce what are now called the Keystone Howards. It manufactured its first railroad watch in 1905--a

Series IX, 16 size, 17 jewel model. All of its railroad watches, produced until 1924, were 16 size, and they were produced in 18 different grades. Howard next produced the Series XI, 21 jewel watch--the railroad chronometer. To quote Townsend again, "All Howard watches are very desirable."

As Table 17 indicates, Waltham Watch Company made thousands of railroad-quality watches. The watches shown in the table total over 600,000, and these do not include some models and grades. Waltham made 65,600 Ball watches, the first being a 16 size AWW Ball model. Figure 45 shows a Waltham Ball in our collection. It is number B 239,571 (made in 1907). It has the standard Ball features: 16 size, 19 jewels, lever set, adjusted to 5 positions. In numbering these watches, the first two numbers were dropped and the letter B was added (presumably for Ball). But the 18 size Waltham was an exception: it kept the full serial number. To further complicate the numbering process, after 1943, 1B and 2B series were produced. Treiman suggests that there may be some duplication of numbers in the Hamilton, Illinois and Waltham production runs, particularly for the B 600,000 and B 800,000 serial numbers. (13)

When the Ball Watch Company introduced a special run of railroad watches for the railroad unions, Waltham participated in their manufacture. It produced 16 and 18 size watches with B of RT, B of LE, B of LF, ORT and ORC on the dials--each standing for a union name (ie, Brotherhood of Railroad Trainmen, Brotherhood of Locomotive Engineers, etc.). But what is more impressive is that these came in different colors--B in blue, R in green and T in red. (14)

Over time, several refinements were made to railroad watches. One was the up and down indicator, which prevented an inadvertent running down of the main spring. A second was the non-magnetic watch. Magnetism, which built up as the watch ran, affected the

Table 17

**RAILROAD WATCHES
PRODUCED IN MASSACHUSETTS**

Name of Company	Size	Number of Jewels	Model	Number Produced
Canadian Pacific Railroad	18	17	1883 1892	2,620
"	16	17	1908	1,950
CRTS	18	17	1892	900
"	16	17	1908*	450
Hampden (Railway)	18	15	1888	?
E. Howard & Co.:				
(Ball)	18	17	--	250
Series 0	16	21-23	--	?
Waltham:				
American Watch Company	18	15	1857	300
Ball	16	16-23	--	56,650
Crescent Street	18	17	1870	1,000
"	18	17-21	1883	3,251
"	18	19-23	1892	52,641
"	16	19-21	1899	19,226
"	16	21	1908	86,090
Railroader	18	17	1892	160
"	16	17	1892 1899	100
Riverside Maximus	16	19-23	1899	500
Road Master	18	17	1892	100
"	16	17	1899	50
Premier	16	23	1908	1,200
642,645	16	17	1908	?
"	16	23	1912	8,000
Vanguard	18	17-23	1892	105,828
"	16	19-23	1899	34,300
"	16-18	19-23	1908	141,650
"	16	21	1621	37,617
"	16	23	1623	59,000
U.S. Watch Co:				
President	18	17	--	?

*
    Townsend states (p. 17) that Hampden made its first railroad watch in 1888--The Railway." He offers a sketch of one with number 57,865 and "Springfield, Mass." on the plates. This must be an error, because this watch was made in 1877 and the company had moved to Canton, Ohio by 1888. See W.F. Meggers, "The Railroad-Marked Watch" NAWCC Bulletin, XXIV (Feb. 1982), p. 6.

Sources: George E. Townsend, <u>American Railroad Watches</u> (1977); Roy Ehrhardt, <u>Waltham Pocket Watch Identification and Price Guide</u> (Kansas City, 1976).

— 196 —

FIGURE 45   *A Waltham Ball Watch*

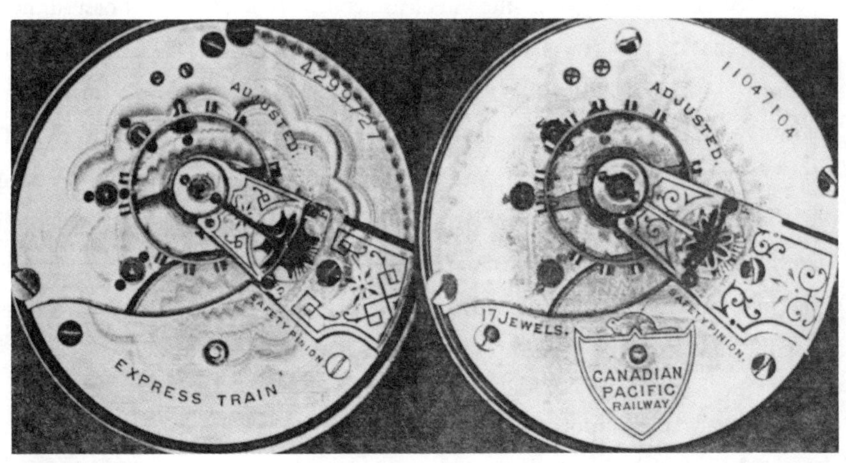

FIGURE 46   *Two 18 size Waltham Model 1883 Railroad Watches.*
The left side features the "Express Train;" the right side is an
Appleton, Tracy & Co. model made for the Canadian Pacific Railway

hairspring and the balance wheel.   About 1920, the Elinvar hairspring was developed. It changed little as temperature varied, and it was affected very little by magnetism. Watches with the new hairspring usually were so designated by the word Elinvar on the pallet cock.

During the 1930's, as the number of watch companies declined, the supplies of railroad watches declined also, so that when World War II commenced, there was a shortage of railroad watches. This was accentuated by the shift in production to war goods, particularly the timers and precision instruments for military uses. During the war, only Hamilton, Elgin and Waltham continued to make railroad watches. Because of the reduced supply, railroads were compelled to reduce their standards for timepieces. But after World War II, Waltham resumed full production of railroad watches. Its new watch was the 16 size, 23 jewel 1623 Vanguard, which was produced until 1954.

Watches for Trolleys

During the 1880's, the electric trolley replaced the horse-drawn railway cars, and in the next decade, the trolley was extended to intercity transportation--this giving birth to the electric interurban railway. These lines grew rapidly, reaching a peak mileage of 18,000 by World War I. (15) The decline of the electric railway was related to the efforts of the automobile manufacturers to stifle the growth and survival of these companies. Eventually, Americans came to prefer the automobile over the electric railway. By 1960, most service was discontinued. (16)

What is important, here, is that the interurban railways established watch inspection systems that were patterned after those of the steam railroads. Because the interurban railways offered frequent service (sometimes on single tracks), they had to rely on accurate timing for "meets"--when trains passed each other.

The interurban lines adopted the system, made common earlier by Webb Ball, of appointing watch inspectors. Interurban motormen and conductors had to present their timepieces to local inspectors at regular intervals. The main question was, what watches should be used by these railway personnel? The Hamilton Watch Company addressed this specific problem; it made several models of "electric specials," in the 926 grade. The Hamilton 948, introduced in 1913, and the 974, first made in 1924, were also intended for use on the electric railways.

The above watches met the standards established by the leading electric railroad companies. These standards included: that the watch should not be smaller than 16 size, that it have no less than 16 jewels, that it be tested in at least 3 positions and that it not vary by more than 15 seconds in these positions.

Waltham produced several watches for this electric railway market. The 18 size, 17 jewel, 1892 Appleton, Tracy model met the above standards, since it was adjusted to 3 positions. When the steam railroads required adjustments to 5 positions, this grade was produced for the electric railways throughout the 1920's.

In 1925, Waltham introduced its 16 size, 21 jewel grade 1621 watch, which was intended for sale to bus drivers and streetcar operators. It was produced in open face, in either pendant or lever set. The 1621 model was made for 30 years (1925- 1955), and it sold for $12 less than the comparable Vanguard grade. (17)

The other Massachusetts company, Howard, chose not to make a special watch for the electric railways. Instead, it was content to sell its regular line to whomever wished to buy--whether it be a steam or electric railwayman.

References

1. See August C. Bolino, *The Development of the American Economy*

(Columbus: Charles E. Merrill, 1966), Chapter 6.

2. Based on a lecture delivered by Carlene E. Stephens at a meeting of Chapter 12 of the National Association of Watch and Clock Collectors, Lanham, Maryland, October 10, 1984.

3. W. F. Meggers, "The Railroad-Marked Watch," NAWCC *Bulletin*, XXIV (Feb. 1982), p. 4.

4. Thomas De Fazio, "Ball & Company and the Vanderbilt Railroads," NAWCC *Bulletin*, (October 1976), p. 411.

5. George E. Townsend, *American Railroad Watches* (1977).

6. Harrison F. and Mary E. Babcock, "The Railroad Watch," NAWCC *Bulletin*, IX (October 1960), p. 385.

7. William E. Miether, "The Babcocks--Mary and Harrison--and Webb C. Ball," NAWCC *Bulletin*, XI (Oct. 1964), pp. 439-446.

8. Lawrence W. Treiman, "Railroad Watches and Time Service," NAWCC *Bulletin*, XV (April 1972), 651.

9. *The New Times*, Jan. 26, 1888, p. 4.

10. R. E. Gould, "The Testing of Timepieces," U. S. Department of Commerce, *Circular C* 432, 1941.

11. Gould, "Testing of Timepieces," p. 1.

12. Townsend, *American Railroad Watches*, p. 24.

13. Lawrence W. Treiman, "Answer Box," NAWCC *Bulletin*, XVIII (April 1976), p. 224.

14. Meggers, "The Railroad-Marked Watch," p. 16.

15. Lawrence W. Treiman, "Timing the Trolleys," NAWCC *Bulletin*, XX (Feb. 1978), 3-19. This section depends heavily on Treiman's excellent article.

16. See especially, George W. Hilton and John T. Due, *The Electric Interurban Railways* (Palo Alto: Stanford University Press, 1960).

17. Treiman states that this watch was introduced "During the
1930's," but this is an obvious error, because the first run was
number 28,932,001 (1925).

Watches that were intended for non-pocket use have been produced for over 300 years. According to Gohl, David Rousseau "made a small verge fusee movement of only 18 millimeters in diameter" in the late 1600's. (This is the size of a United States dime.) (1) Rousseau's watch is currently exhibited in the Dresden Museum. Since Rousseau's time, many nations have claimed the "smallest watch" prize. In England, George III acquired a 10 millimeter repeating watch from watchmaker John Arnold of London (about 1764), and in Geneva, about a century later, Edoard Sordet made 3 tiny key-wind watches, measuring only 9 millimeters in diameter. One is now the property of the Horological Museum in that city. In the next century, watches got even smaller. According to J. Bunzel, of Maiden Lane, New York, he imported the smallest watch in the world in 1888 for 280 francs. It was 19/32nds of an inch in diameter, full-jeweled, gilt, cyclinder, fully regulated and stem wind. At the 1896 Geneva Exhibit, Paul Ditisheim, of La Chaux-de-Fonds, showed a 6.75 millimeter stem wind watch, which weighed less than a gram!

Mass Producing Small Watches

Small watches are an interesting horological item for museums, but our interest is wrist watches that were mass produced, especially those made in Massachusetts. In this case, too, the history begins in Europe. The original wrist watches were called "bracelet watches." Naturally, these were attractive to the ladies, and as such, the French constituted a part of the early demand for this type of watch. The Empress Josephine in 1806 wore two watches to show both time and date.

The impetus for mass-producing wrist watches came from the military. A German naval officer, who had great difficulty holding a watch during a war maneuver in the 1880's, recommended that if a watch was attached to the wrist, it would free the use of both hands in battle. As a result of this recommendation, Swiss watch makers received orders for small watches for German naval officers. The Swiss attempted to introduce these watches in the United States, but the initial efforts failed. Women considered them jewelry and men rejected them completely.

Arthur Tremayne offers another view of the beginning of wrist watch production. He states, "The first of which I have positive news was worn by an officer in the South African war. This soldier was in private life a jeweller, and seeking a means of carrying a watch he took a 12 size watch and removed the bow. He took the case to the late W. E. Tucker of H. Williamson, Ltd. of London, who had loops soldered to the case which was then sent to E. J. Pearson, a saddler, of Clerkenwell to make straps to retain the watch to the wrist. This was in 1900." (2) The Swiss took up the idea, and by 1906 they were exporting wrist watches to the United States.

American production of wrist watches commenced in the first decade of the twentieth cntury. By this time, wrist watches had been accepted throughout the world. The Dueber Watch Case Company had produced wrist watch cases earlier, but it made its first complete watch in 1911. While the Waltham Watch Company appeared to be ignoring wrist watch production in 1911, its 6/0 size movement, first produced in 1901 as an 1898 model (number 10,000,000), was to become the backbone of military wrist watch output for World War II. The 1898 6/0 size movement was replaced by the 1912 movement, which in turn was replaced by the 1942 movement. The 1898 and 1912 "Jewel Series" were produced in hunting and open-faced models respectively. The first 1912 movement was numbered 18,059,001. (3) Waltham also produced the 0 size, which was about the size of a quarter, and which was suitable for a ladies lapel pin or chatelaine.

But the problem was that men still preferred pocket watches: 16 and 14 size in the East and 18 size in the West. (4)

FIGURE 47 *The 6/0, 1898 model "Ruby" wrist watch movement number 26,710,943 (1928)*

By the beginning of World War I, wrist watches were winning acceptance by American women, and nearly all American watchmakers had introduced these "wristlets." Hamilton had 2 models by 1912 and Waltham introduced its first complete wrist watch by 1914. But the Swiss, who had pioneered these watches, dominated the market: Gruen had 26 models by 1912. And while the American men still considered wrist watches effeminate, Gruen offered 7 military wrist watch models in 1918.

To make wrist watches more acceptable to men, the watch dials were made bolder--with larger hands and luminous numbers. The Waltham catalog for 1919 showed that in 5 years since it first introduced its first complete wrist watch, the company was offering several models for women but only one "strap watch" for men. It was sold in 7 and 15 jewels and with silver or nickel case. The catalog stressed that, "It is an essential of every-day life."

Between 1907 and 1920, most wrist watches had round movements. For this reason, jewelers could easily place a round movement into a wrist watch case. This was particularly true of the 0 size--which was usually worn by women as a neck piece.

A Digression on Sizes

At this point, since so many wrist watches are measured in lignes, we need to clarify the relationship between inches and lignes. As we stated in Chapter 2, when Aaron Dennison was designing watches for the Boston Watch Company, he devised a new system for sizing watch movements, which set one inch equal to 0 and designated each additional number as 1/30th of an inch. Thus, a 16 size watch was one 16/30ths inches in diameter (measured across the back plate). Since the watch was to be placed in the case from the front, the front diameter had to be slightly larger (the "drop"). Dennison arrived at his new system as a necessary requirement for making interchangeable parts.

The American system was revised (probably by Dennison) sometime soon after 1857. The size of the back plate was increased by 2/30ths of an inch. An 18 size was then measured as one and 20/30ths, but the back plate remained at one and 23/30ths inches. All sizes smaller than 18 size were reduced accordingly. (5) By contrast, the French or Swiss ligne measurement of watches is based on the Paris inch, which is 1/12th of a Paris foot, which is 1/6th of the French toise. The Paris inch has 12 lignes.

When we turn to wrist watch sizes, the measurements are peculiar. The 0 size became 1/0, and the 2/0 was only 1/30ths of an inch smaller than the 0 size (now 1/0). By this reckoning, the 10/0 size was only 9/30ths of an inch smaller than the 0 size, across the front plate. To find the size of a wrist watch, the diameter of the front plate equals one and 5/30ths minus one less than the number before the /0. For example, a 6/0 movement equals one and 5/30ths minus 6-5/30, which equals 0. But more puzzling, while the large size watches had a constant drop (about 3/30ths of an inch), for wrist watches, the drop decreased as did the size.

Table 18 offers a comparison of the American and European systems. We see there that the Paris inch (12 lignes) equals 1.066 American inches, and that the American 0 size equals 1.167 American inches. It should be noted also, that often when American companies used grade numbers for wrist watches that these were in lignes. For example, the Waltham 400 and 670 models were 4 lignes and and 6 and 3/4 lignes.

The Wrist Watch in its Dominant Period

In the 1920's, American watch companies had to supply watches for the Flapper era--the time of the Jazz Age, the Charleston and Art Deco. They responded with a very large variety of watch shapes, including rectangular, oval and tonneau. The Art Deco Exhibit in Paris in 1925 encouraged fancy designs. As Gohl points out, the "Bracelet Watch Crystals" catalogue for 1923 listed 148 shapes and sizes. (6)

In 1927, for the first time, wrist watch production exceeded that of pocket watches, and thereafter pocket watches captured a declining segment of the total market. How was Waltham faring in this new situation? As Table 19 indicates, between 1910 and 1930, Waltham made large numbers of 10 ligne and 7 and 1/4 ligne watches. The Table also shows that the 5 and 1/4 ligne, 9 ligne, 7 and 1/2 ligne and bracelet watches were relatively scarce. Only 2,100 9 ligne watches were made, mostly Patrician, 17 jewel movements; and only 3,500 5 and 1/4 ligne, mostly 15 jewel (although some 7 and 17

Table 18

AMERICAN AND SWISS WATCH SIZES COMPARED

American Watch sizes	Diameter of pillar plate			
	Lignes	mm	Inches	30ths of an inch
19	——	45.72	1.800	54
——	20	45.12	1.776	——
18	——	44.87	1.767	53
17	——	44.03	1.733	52
16	——	43.18	1.700	51
——	19	42.86	1.687	——
15	——	42.33	1.667	50
14	——	41.49	1.633	49
13	——	40.64	1.600	48
——	18	40.61	1.599	——
12	——	39.79	1.567	47
11	——	38.95	1.533	46
——	17	38.35	1.510	——
10	——	38.10	1.500	45
9	——	37.25	1.467	44
8	——	36.41	1.433	43
——	16	36.09	1.421	——
7	——	35.56	1.400	42
6	——	34.71	1.367	41
5	——	33.87	1.333	40
——	15	33.84	1.332	——
4	——	33.02	1.300	39
3	——	32.17	1.267	38
——	14	31.58	1.243	——
2	——	31.33	1.233	37
1	——	30.48	1.200	36
0	——	29.63	1.167	35
——	12	29.33	1.155	——
2/0	——	28.79	1.133	34
3/0	——	27.94	1.100	33
4/0	——	27.09	1.067	32
——	12	27.07	1.066	——
5/0	——	26.25	1.033	31
——	11 1/2	25.94	1.021	——
6/0	——	25.40	1.000	30

American Watch sizes	Diameter of pillar platoe			
	Lignes	mm	Inches	30ths of an inch
——	11	24.81	0.977	——
7/0	——	24.55	.967	29
8/0	——	23.71	.933	28
——	10 1/2	23.69	.933	——
9/0	——	22.86	.900	27
——	10	22.56	.888	——
10/0	——	22.01	.867	26
——	9 1/2	21.43	.844	——
11/0	——	21.17	.833	25
12/0	——	20.32	.800	24
——	9	20.30	.799	——
13/0	——	19.47	.767	23
——	8 1/2	19.18	.755	——
14/0	——	18.63	.733	22
——	8	18.05	.711	——
15/0	——	17.78	.700	21
16/0	——	16.93	.667	20
——	7 1/2	16.92	.666	——
17/0	——	16.09	.633	19
——	7	15.79	.622	——
18/0	——	15.24	.600	18
——	6 1/2	14.66	.577	——
19/0	——	14.39	.567	17
20/0	——	13.55	.533	16
——	6	13.54	.533	——
21/0	——	12.70	.500	15
——	5 1/2	12.41	.488	——
22/0	——	11.85	.467	14
——	5	11.28	.444	——
23/0	——	11.01	.433	13
24/0	——	10.16	.400	12
25/0	——	9.31	.367	11
——	4	9.02	.355	——

Source: R.E. Gould, "The Testing of Timepieces," U.S. Department of Commerce, Circular C 432, 1941, pp. 21 – 22.

Table 19

WALTHAM WRIST WATCH PRODUCTION
(1910 – 1930)

Size	Grades	Jewels	Number Produced
10 Ligne	Maximus Lady Waltham 461 465	15–17	271,500
7 1/4 Ligne	Rectangular 761 767 765	15	180,500
6 1/2 Ligne	Oval 665 667	17	42,000
7 1/2 Ligne	Patrician Lady Waltham Waltham Rectangular	17	8,700
Waltham	Bracelet	15–17	8,500
5 1/4 Ligne	Rectangular	7–17	3,500
9 Ligne	Patrician	17	2,100

Source: Roy Ehrhardt, _Waltham Pocket Watch Identification and Price Guide_ (1976).

jewel models were produced). Almost the same number of 7 and 1/2
ligne and bracelet watches were made: 8,500 bracelet and 8,700 7
and 1/2 ligne. The bracelet watches were made in 1917 only and in
15 and 17 jewels.

The stock market crash and the Great Depression brought down
a number of watch companies too. As we related in Chapter 2,
between 1929 and 1931, the New York, Hampden, Illinois, Howard
and South Bend watch companies were bankrupted or sold out to
competitors. The demise of the pocket watch contributed to these
failures. In the 1930's, new wrist watch models were the thing.
Waltham brought out its 400 and 450 models--the 400 baguette model
being the smallest watch that was ever mass produced in the United
States. It measured only 9 millimeters by 20 millimeters. It also
had a unique fifth wheel between the fourth wheel and the escape
wheel, which made it beat faster (1/6 beat, or 21,600 vibrations per
hour compared to the normal 1/5 beat of 18,000).

Gohl states that the 400 was introduced in 1932 and made until
1938, but according to Ehrhardt, the first production run of the 400
was in May 1931 (27,645,001- 7,000), and the last run was numbers
29,929,001-39,000, in June 1939. (7) The 450 model, Grade 455,
had 15 jewels and was produced in approximately the same time and
production runs as the 400.

Some persons believe that the 400 model was, in fact, a Swiss
watch. Myron Everts, who worked on the movement, and who
ordered parts for it, states, "My guess is that the whole movement
was assembled in Switzerland." But Joseph Dean, who worked at
Waltham, responded that all parts were American, with the possible
exception of the jewels. For him, the movement followed "the pattern
of Waltham workmanship." And, as he stressed, Waltham did not
begin to import movements until much later. (8)

THE "450"

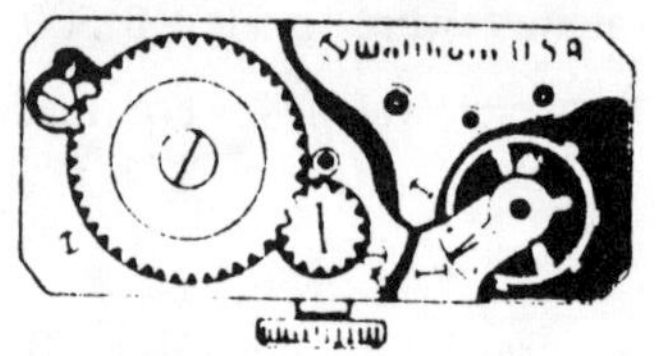

MOVEMENT No. 455

15 Jewels

FIGURE 48   *The Waltham "450."   Taken from the E. & J. Swigart Co. Catalog*

The 650 (6 and 1/2 ligne) oval wrist watch was first produced in 1929 (26,469,001) and was made regularly until August 1942. Other watches, which were first made in the 1930's, were also discontinued during World War II.   These include the 870 (8 and 3/4 ligne), first made in 1936 (28,966,001) and discontinued in August 1942, and the 670 round movement with 17 jewels, first manufactured in May 1938 and halted in March 1941.

Wrist Watches in World War II

Wrist watches have been issued to military personnel since 1912. (9) The standard issue watches in World War II included the Elgin 8/0 size, 15 jewels; Elgin model 987A, 17 jewels; and the Waltham 6/0 size, also 17 jewels.   The Waltham 6/0 size wrist watch that was used as a service timepiece came in two basic models: the 9 jewel, designated by the prefix OC, and the 17 jewel, with OD before the serial number.   The 17 jewel model was usually issued to the Air Corps. (10) The "hack watch" was a special type of 6/0 watch.   It had 16 jewels and a sweep second hand that allowed the navigator on the airplane to reset the watch for particular duties (for example, to calibrate instruments or to rendevous to a certain geographical point). The first of these hack watches was made in August 1942 and the last

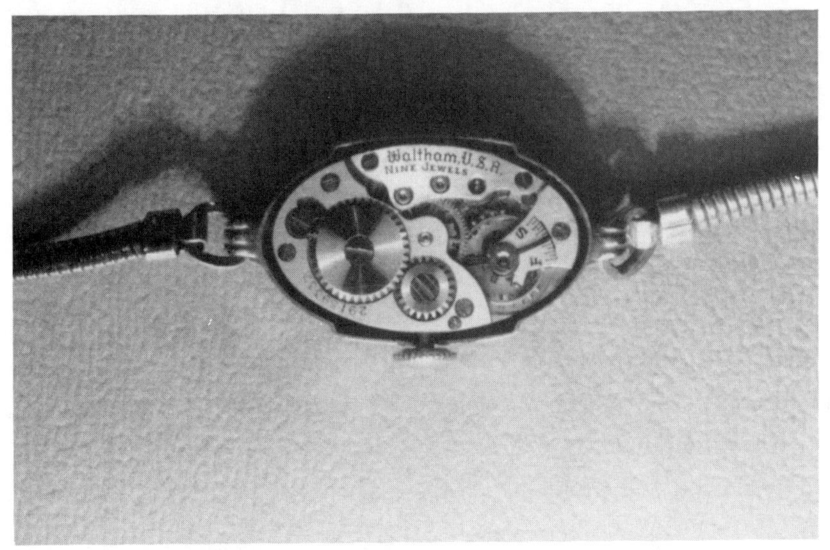

FIGURE 49 *A Waltham 650 movement. This 9 jewel, Grade 659 is number 29,159,337 (1937) and is adjusted*

one in June 1945.

The 7 and 1/2 ligne watch, which was introduced in 1915 (20,074,501), was replaced by the 7 and 1/4 ligne watch in 1924. The 750 model was re-introduced in 1941 as model 767, with a tonneau shape. During World War II, it was produced with 9, 17 and 21 jewels.

Once the American factories were geared up for full production, there were enough wrist watches for military purposes, but the civilian supply became extremely short. In 1944, the War Production Board listed wrist watches among those with serious shortages, along with refrigerators, sewing machines, electric ranges and washing machines. (11)

Wrist Watches to the End

The Waltham post World War II wrist watch production is

shown in Table 20. It is clear from this table that much of the output after 1945 was not new; several later models of earlier watches were developed (ie the 750B and the 6/0D). In this connection, we need to point out a misleading aspect in Ehrhardt's Waltham volume. On page 168, he presents photographs of Waltham watches that he labels, "1948 WALTHAM WRIST AND BRACELET WATCH MOVEMENTS." He pictures 19 movements. In fact, on pages 132 and 133 of the same volume, he lists only 3 wrist watches that were produced in 1948: the 6/0C, the 750B and the 678 models. The production of the others was long discontinued. The 5 and 1/4 ligne model was terminated in 1923; the 10 ligne in 1930; the 400 in 1939; and the 650 in August 1942.

FIGURE 50 *Some Waltham wrist watches*

Several wrist watches were approved for use as railroad time pieces after 1960. These include the Elgin 13/0 size. 23 jewel, B. W. Raymond; the Ball 13 ligne. 21 jewel Official RR Standard; the Bulova Accutron (models 214 and 218). Later. the quartz and electric models were also approved (Accutron 201 and 202 models).

Table 20

WALTHAM WRIST WATCH PRODUCTION
(1945 - 1953)

Model	Production Dates	Style	Number of Jewels	Position
750 B	August 1942–March 1953	Barrel	17–21	4 adj.
675	June 1945–June 1948	Barrel	17–21	4 adj.
678	May 1946–April 1952	Barrel	17–21	4 adj.
6/0 C	August 1946–June 1949	Round	17–21	Pos.
6/0 D	August 1946–August 1953	Round	17–21	Pos.
870 R	December 1946–November 1947	Round	17–21	4 adj.
679	July 1949–January 1950	Barrel	19	Pos.

Source: E & J Swigart, Illustrated Manual of American Watch Movements (Cincinnati, 1953); Roy Ehrhardt, Waltham Pocket Watch Identification and Price Guide (1976).

What is significant about the above list is that it contains no Waltham watches. This is indicative of the malaise that we discussed in Chapter 9. The Waltham answer to its problems was to import watch movements. Figure 50 pictures some of these movements. They were distributed by Waltham under its own name and model numbers. But parts had to be ordered from the Swiss identification system. *Sic transit gloria mundi.*

References

1. Anthony Gohl, "The Wrist Watch," NAWCC *Bulletin*, XIX (December 1977), p. 586. I am indebted to Gohl for providing the framework for this chapter on wrist watches. Although more has been written since his article, as he said then, "this phase of horology . . .has had little attention and in many ways is still been overlooked."

2. Arthur Tremayne, "The Introduction of Wrist Watches," NAWCC *Bulletin*, II (Aug. 1947), p. 265.

3. E & J Swigart Co., *Illustrated Manual of American Watch Movements* (Cincinnati, 1953), p. 146.

4. Earnest A. Cramer, "The Watch Case and the Watch," NAWCC *Bulletin*, VI (Feb. 1957), 410.

5. Mathews, A. E., "American Watch Sizes," NAWCC *Bulletin*. XIV (Aug. 1970), p. 434.

6. Gohl, "The Wrist Watch," p. 592.

7. Roy Ehrhardt, *Waltham Pocket Watch Identification and Price Indicator* (Kansas City, 1976), pp.127-29.

8. See the "Vox Temporis" section of the NAWCC *Bulletin*, IX (Feb. 1960), p. 175.

9. Williamson, R. C. "Military Timekeepers," NAWCC *Bulletin*, (April 1970), p. 257.

10. U. S. War Department, "Wrist Watches, Pocket Watches, Stop Watches, and Clocks," *Technical Manual* 9-1575, April 6, 1945, p. 161.

11. *The New York Times,* June 22, 1944, p. 25.

12. See, for example, Circular Notice No. 523-78, of the Louisville and Nashville Railroad, Oct. 13, 1978 and Robert G. Williams, "Some Notes on Modern Railway Timekeeping," NAWCC *Bulletin,* XVI (Feb. 1973), p. 980.

CHAPTER 13:
IMITATING MASSACHUSETTS

Imitating Key Winds

As American watches gained in quality and stature, there was a greater incentive for Europeans to make imitations. These imitations took a large variety of forms: some were complete watches with slightly misspelled names; others had correct names (of American companies) but had European features; and still others had imitation movements in genuine American cases.

The large majority of these watches were Swiss, and so they are often called "Swiss Fakes." The Swiss turned to deception, because by the mid-1870's, the supremacy of watch production had shifted from Europe to the United States. As we told in Chapter 9, Royal Robbins set high standards and low prices as his goals for the Waltham Watch Company. By the time of the American Centennial, the United States had a clear lead in watch technology and production. In fact, the Swiss must have sensed their imminent decline, because they began to produce imitations fairly early in United States watch history. These fakes reached a peak production in the 1880's, hence many were key winds.

Figure 51 shows one of these early key wind imitations-- a Massasoit. It looks very much like a William Ellery model Waltham, but on closer examination we find a Swiss cap jewel and European escape wheel and pallet. But the Massasoit is confounding, because some movements with that name were made by the Dennison, Howard Company around 1854. They can be distinguished by their unique dials, which were made for the Turkish trade. We wonder, then, whether any other Massasoit movements found their way into the American markets. If they did, they may not be Swiss fakes. (1)

Because some imitation watches were of very high quality and

— 216 —

FIGURE 51 *A Massasoit*

not so easily detected by novices, the American watch companies pressured the Congress to act, which it did in the legislation of March 3, 1871. The law prohibited the importation of movements which were engraved with false names. (2) This law reduced the flow of fakes temporarily, but soon the Swiss were able to improve the quality of these copies, hence they were not so easily detectable. They made the movements conform to the United States standards (ie in the size of the cap jewels, in the dial feet and in the dial assembly). However, they had difficulty with serial numbers, and they could not duplicate easily the American gilding process. And some watches were 18 size with cylinder movements--an impossible combination. Sears Roebuck, which we know began as a watch retailing company, reacted honestly to the problem of Swiss fakes. Its' 1897 catalog featured several key wind watches with cyclinder movements. The catalog states, "They will keep very good time, but we do not warrant them. Do not be deceived by this class of goods. We sell them for what they are." (3) The catalog states that the watches were made by poor peasants in the mountains of Switzerland during the winter months, and that they receive 15 cents per day for their work. For comparison, in this time, the Americans earned one dollar per day.

Although the 1871 law forbade the importation of these watches, the supply continued, because many (or most) movements were smuggled into the United States. Hauptman relates a story from the *Detroit Tribune* of February 12, 1873, which was duplicated many times. In the story, J. W. Pierson, of the E. Howard Watch Company, was offered "Howard" watches at $10 to $20 less than the regular price. When he investigated, he saw that they were imitations, so he called in the Collector of Customs, who confiscated the entire shipment. Hauptman claims that, "By 1876 (the problem) was virtually non-existent." Surely, this is an exaggeration of the extent to which the smuggling of fakes was reduced then or anytime after. (4) Smuggling went on in good years and in depressions. *The New York Times*, for example, reports that in 1934, there was a "nation-wide racket in dials" to make fakes resemble nationally-known brands. In this case, detectives posed as engravers in the small factory at 43 Lexington Avenue, New York. This forgery involved 2,500,000 watches. (5)

Characteristics of Swiss Fakes

In the matter of these fakes, how shall we know them? The answer is that they have a set of common characteristics.

1. The names were misspelled (more later). 2. The dials had 2 feet instead of the more common 3. 3. The gilding process was not uniform, hence color differences were noticeable. 4. The Balance wheels were often solid and some had timing screws. The American balance wheels were cut and were of different metals. 5. The engraving on Swiss watches was less fine than on American plates. The roughness is detectable in the non- straight lines. 6. The word "Swiss" was usually placed in some inconspicuous place (under plates,

wheels or even hidden in the scroll work of the engraving).   7. The cap jewels on the imitations were larger and clumsier than the comparable American part.

In commenting on Hauptman's article, Pease presents two examples from his own collection that illustrate some of the above points.   The first looks like a standard American Watch Company key wind 57 model, but it has a winding mechanism that would "feel at home" only in a 16 size three-quarter plate or bridge model.   And what is almost comical, it is labeled "T. G. American Watch, New York." (!)   The other example also looks like like a 57 model.   But the name is in block lettering, not script, and it reads "London Watch Company."   Pease wonders why this very good imitation gives itself away so easily when it might not otherwise have been detected. (6)

Imitation Railroad Watches

As economists say, *caveat emptor* (let the buyer beware), because there are some very good imitation railroad watches that can fool the novice.   While most of the fakes were Swiss, some were made by American companies (ie Seth Thomas and New York Standard).   These watches varied widely.   Some had 21 or 23 jewels; some were fully adjusted or had 3, 5, 6, or 8 adjustments; some advertised double rollers, non-magnetic or "Special" and many carried numbers that were close to American models (ie 543 or 99).   Also, many had locomotives or the word railroad or a company name on the plates.

The example shown in Figure 52 is typical.   The Time Ball Special has a 60-minute railroad dial, 21 jewels, 3 adjustments and was made by Morthier-Sandoz.

A List of Fakes

As we stated above, the Swiss attempt to deceive American consumers was a clumsy effort.   The names that were chosen were

Table 21

IMITATING MASSACHUSETTS WATCHES

Name of Company	Name on Plate
The American Watch	The Barton
Attleboro Watch Co.	Boston
B.F. Bartlett	Boston
B.F. Barrlett	Waldham, Mass.
P.F. Barzlet	Waldham
P.G. Barzlet	Waldham, Mass.
P.J. Barzlet	Waldham, Mass.
P.S. Barrett	Waltham, Mass.
P.S. Bartlets	Waldham, Mass.
P.S. Barzlet	Waltham, Mass.
S.P. Bartley	Boston
P.T. Barzlet	Waldham, Mass.
P.T. Barslet	Waltham, Mass.
California Watch Companies	Boston, Mass.
Great Eastern Watch	Rosebury, Mass
Imperial Watch Co.	Boston, Mass.
Marvin Watch Co.	Springfield
Massasoit Watch Co.	Boston
National Watch Co.	Boston, Mass.
Rosebury Watch Co.	Boston, Mass.
Roxbury Watch Co.	Boston, Mass.
Union Ellery	Boston, Mass.
U.S. Watch Co.	Boston, Mass.
Wallingford Watch Co.	Wallingford, Mass.
Wm. Elleby	Boston, Mass
Wm. Ellerty	----
Wm. Elley	----

Source: B.J. Jackson, "Imitation American Key Wind Watches," NAWCC Bulletin, XXI (August 1979), 426-433; James W. Gibbs, "Sucker Bait--- (Horological, not Piscatorial)," NAWCC Bulletin, IX (August 1961), 791-798; Roy Ehrhardt, Waltham Pocket Watch Identification and Price Indicator (1976).

FIGURE 52   *The "Time Ball Special"*

almost bizarre.   Can Americans really have been fooled by such variations on the Waltham theme as are shown in Table 21?   There were at least 11 variations of P. S. Bartlett.   We are left, then, with the question raised by Gibbs, "The curious thing again is why these foreign watches were brought in with domestic names on them after the Civil War period when the American watch system seemed to be working well." (7)

References

1. Charles Kalish, "My Favorite Watches," NAWCC *Bulletin*, IX (Dec. 1960), p. 488.

2. Wesley R. Hauptman, "Swiss Imitations of Early American Watches," NAWCC *Bulletin*, IX (Aug. 1960), p. 429.

3. Quoted in B. J. Jackson, "Imitation Key Wind Watches," NAWCC *Bulletin*, XXI (Aug. 1979), p. 430.

4. Hauptman, "Swiss Imitations," p. 273.

5. *The New York Times*, Oct. 19, 1934, p. 46.

6. Hamilton E. Pease, "Imitations of American Watches," NAWCC *Bulletin*, IX (Apr. 1961), 709-710.

7. James W. Gibbs, "Sucker Bait---(Horological, not Piscatorial)," NAWCC *Bulletin*, XIX (Oct. 1977), p. 792.

In this chapter, we cover what you might call the rest of the watch--the cases, dials and hands.

The Watch Case

Early European watch cases ran a very wide gamut of sizes, colors, metals, designs and processes. They were ornate and they were plain. Some were enameled (using ancient techniques) and others were "engine turned."

Our interest is in American watch cases. Its history paralleled the development of and the growth of watch movements, which we have described in earlier chapters. Before 1850, watch case production involved a considerable amount of hand labor. The modern American watch industry seems to have begun in Philadelphia. Cramer indicates that the 1850 Philadelphia city directory lists 18 shops manufacturing and repairing watch cases. One of these was Eliashib Tracy of Tracy, Baker and Appleton, Tracy fame. Cramer tells us that by 1890, there were 47 companies in Philadelphia turning out 6,000 cases per day, using 5,000 operators. (1)

The true beginning of American case production came in 1851, when two brothers, Reese and Randolph Peters, started making cases in an old building at Dock and Walnut Streets. They made mostly silver cases, and they made them by hand. Working with the Peters brothers was James Boss, who like them, was experimenting with making cases with low gold content. When all of the early combinations of metals tarnished badly, he began working on an idea provided by Reese Peters--to find a gold surface with a base metal center. Boss's solution was to combine two plates of gold with a central piece of base metal, without the aid of solder. This was accomplished by making the gold so smooth that the surfaces would adhere by suction. This process was the original "gold filled" plate.

which was replaced later by a less expensive process.   The Boss
process was patented on May 3, 1859; it guaranteed the case to wear
20 years, and the cases so stated.   In fact, Boss offered a money-back
guarantee.   (2)

After the Civil War, specialization in watch manufacturing
became common--some companies made just movements and others
cases, and it was up to the jeweler to fit one to the other.

FIGURE 53   *A J. Boss Case.   This swingout, 25-year
case houses a 19 jewel Waltham Riverside movement,
number 12,560,592 (1903).*

The number of watch case manufacturers grew rapidly.   The 18
shops that functioned in 1850 produced 285 cases per day; by 1929,
the watch case industry could produce 8,300,000 cases. (3) Although
Boss dominated the industry early, in time, he had many competitors.
In 1871, Boss sold out to John G. Stuckert, also of Philadelphia.
When he died, the business was sold to Hagstoz & Co.   Hagstoz
hired Charles M. Thorp as business manager of the firm, and

together they became quite famous with their exhibit at the Philadelphia Centennial Exhibition of 1876. This company was to become a part of the Keystone Watch Case Manufactory at Riverside, New Jersey, and later it was one of several companies that Theophilus Zerbrugg assembled. (4) In 1927, the Keystone Watch Case Company controlled the Howard Watch Company, the New York Standard Watch Company, the Crescent Watch Case Company and the Philadelphia Watch Case Company.

The Crescent Watch Case Company was founded at Chicago in 1882. After an initial success in that city, the company moved to Brooklyn, New York and had a much greater succcess with its star and crescent trade mark.

H. A. Wadsworth & Company also grew rapidly from its small beginnings in Newport, Kentucky in 1889. Harry Wadsworth worked to perfect his process with his two sons Arthur and Randolph. They moved their plant to Dayton, Kentucky in 1900 and eventually became a part of the Elgin National Watch Company in 1953. At its peak, the company employed 1,350 persons and had sales of $16 million. (5)

The demand for cases from Massachusetts watch companies was satisfied at home and abroad. As always, Aaron Dennison was involved. After selling out to Royal Robbins, he went to England in 1870 to sell watchmaking machinery to a company which later became the English Watch Company. Since he had difficulty establishing a movement factory in England, he turned to producing cases. In 1874, he organized the firm of Dennison, Wigly & Company at Handsworth, near Birmingham. The company produced gold filled cases, which the English did not manufacture. They were of various grades and guarantees. For example, the Moon grade of 10 carat gold filled was guaranteed to last 20 years, and as Aked stated, many "do not show base metal after half a century of use if treated properly." (6) In achieving these spectacular results, Dennison used the same mass-production techniques he had learned in Massachusetts, and his cases were of such high quality and low price that Dennison was able to corner the market for cases for Waltham watches in Europe. "The combination of Waltham movements and Dennison cases became a popular one in England and the London Office (of Waltham) prospered." (7)

FIGURE 54 *An Interesting Case Made in Massachusetts*

Waltham produced its own cases for domestic use during most of the 19th century, but in 1890, The Crescent Watch Case Company purchased the entire case business of the Waltham company. This is reflected in Table 22, which shows that the number of cases produced at Waltham grew steadily, until the peak of 242,519 in 1889. In 1890, Waltham sold its silver case business and in the following year its gold business. But in 1903, Waltham and Elgin joined with the Keystone Watch case Company and the Crescent Watch Case Company to form a syndicate to purchase the American Watch Case Company, of Toronto, Canada. This action was taken to gain an entry into the

Table 22

WALTHAM CASE PRODUCTION
(Feb. 1 - Jan. 31)

Number of Cases		Number of Cases	
1859	6,146	1880	55,603
1860	3,768	1881	55,522
1861	1,654	1882	66,965
1862	8,450	1883	82,047
1863	12,059	1884	185,670
1864	14,485	1885	139,047
1865	16,919	1886	191,815
1866	18,714	1887	170,524
1867	23,167	1888	226,918
1868	13,557	1889	242,519
1869	33,038		
1859–1869	**151,957**	**1880–1889**	**1,416,630**
1870	37,449	1890	193,865
1871	42,785	1891	31,808
1872	44,169		
1873	48,285		
1874	31,322		
1875	33,816		
1876	39,543		
1877	38,368		
1878	43,668		
1879	39,888		
1870–1879	**399,293**	**1890–1898**	**225,673**

Grand Total, 1859–1899   2,193,553

<u>Notes:</u>

1. In 1884, for the first time, case production was broken down into silver and gold components. Until then presumably, most cases were silver.

2. All cases produced in 1891 were gold. The Waltham Silver Case business was sold in 1890 and the gold in 1891.

Source: Vernon M. Hawkins (Transcriber), <u>Annual Reports to Stockholders, American Waltham Watch Company, 1859–1899</u> (West Boxford, 1984).

— 227 —

Canadian market, where the duty on watch cases was 30 percent. (8) By 1906, Waltham and Elgin owned 58 percent of the American Watch Case Company. Nevertheless, by 1925, only 15 percent of Waltham movements were sold with cases.

The early gold-filled cases had 20 year guarantees, but in time there was an erosion of quality, making the guarantee meaningless. The manufacturers put all sorts of marks, stamps and other designations to confuse the consumers. In June 1923, the Federal Trade Commission ruled that manufacturers must indicate the fineness of metal used in gold- filled cases. (9)

Making Cases in Massachusett

In Chapter 5, we related the story of John C. Dueber, who began as an apprentice in the shop of casemaker Frank Doll in Cincinnati, Ohio. He built his own case company in Newport, Kentucky, and after merging with the Hampden Watch Company, he moved to Canton, Ohio in 1889. His company produced a full line of cases including gold, gold filled and silver types.

Bates & Bacon was another Massachusetts case company. Located in Attleboro, it sold cases under the "B & B" trademark. At its peak, it employed 200 persons. (10) Bates & Bacon sold both solid gold and gold filled movements for the trade. One of its more famous labels was the "Royal." We should note, too, that the B & B Company was also owned by Keystone. These mergers brought up the question of trusts and of possible antitrust action against the National Association of Jobbers, which dealt in watch cases.

The case department at the Waltham Watch Company was managed by Daniel O'Hara (who achieved more fame as a dial maker). He oversaw the production of the four main parts of the watch case: the center, the back, the cap and the bezel. At Waltham, the cap, bezel and back were attached to the center by as many as

150 operations. All cases there were silver (the gold ones were made in New York City). The silver bars were first rolled into shape, then they were turned, jointed and soldered. After the cap was put on, the crown and the case spring were attached. The case was then disassembled for polishing. In 1884, the case department could produce 650 silver cases per day, using 400 employees. (11)

Table 23 presents a list of casemakers from Massachusetts. It is obvious from the list that very few cases were made in Massachusetts in the 20th century.

Making Dials

The knowledge of making dials was imported into the United States from England and Switzerland. By 1860, many Englishmen from Liverpool, Lancashire and Coventry found employment in the new Waltham Watch Company. John Webb, Jr. was one of them. He was the son of John Webb, Sr., who made a locket for Queen Victoria, that had the Lord's Prayer painted on it that was "the size of a common house fly's wing." At the age of fourteen (1836), Young John was indentured for seven years to a master enameler of Coventry by the Christ's Hospital (a boarding school for underprivileged children). He became a freeman in 1846, married in 1847 and emigrated to the United States in 1855. He started out for Illinois, and he got there by way of the Port of Boston. He must have heard of the new Howard and Dennison factory, because he moved to Waltham, where he became head of the dial department.

In a letter, John Webb, Jr. disclosed that his father had developed the process for making the sunk portion of a watch dial. Because his apprenticeship agreement had sworn him "his secrets (to) keep," he made these dials for Waltham at his home. Obviously, the secret got out. When Webb transferred to the Elgin company, the process was common knowledge to most dialmakers. (12)

Table 23

THE CASE MAKERS OF MASSACHUSETTS

Name of Company	Address	Dates of Production
Bates and Bacon	Attleboro	1883-1890
Bay State Watch Case Co.	Boston	1889
Boston Watch Co.	Boston	1880
Friedli, J. and Co.	Boston	1872
George W. Ladd	Boston	1873
Goddard Watch Co.	Shrewsbury	1809
Langdon Watch Co.	Boston	1857-1862
Langdon, William G.	Boston	1862
Marget Bros.	Boston	1863-1881
Owen and Pound	Waltham	1857-1861
Robert, A.A. and Foster	Boston	1920
Rogers, Langdon and Went	Boston	1852-1857
Serex and Desmaison	Boston	1876-1880
Serex and Maitre Bros.	Boston	1869-1876
Serex and Robert	Boston	1880-1889
Thiery Watch Case Co.	Boston	1876-1890
Waltham Watch Co.	Waltham	1857-1957

Source: Cooksey Shugart, _The Complete Guide to American Pocket Watches_ (Cleveland: Overstreet, 1981), pp. 65-70; _Boston Business Directory_, Various years.

In the 1870's, the making of dials was speeded up by specialization, but the hand process was still too slow for the growth of movement production. Waltham attempted to perfect a photographic process, but it turned out imperfect dials that had to be retouched by hand, and it could only produce one-color dials. (13)

Many of the early key wind watches had what Mathews calls "universal watch dials," particularly for 18 size watches. They were plain, without name. This was true of "all Waltham 18-size, key wind watches, excepting the 'chronodrometer' and the 'Crescent Street' models." (14) These universal dials fit some Elgin watches and watches made by Howard & Davis and the Boston Watch Company, Newark, Cornell, California, Tremont, Melrose and the United States Watch Company movements. These early dials had holes through the feet, through which small pins were placed to hold the dial in place. These "universal" dials had one dial foot near the 3 o'clock position, which interfered with the winding mechanism on hunting case watches. In those situations, the 3 o'clock dial foot was often removed.

If one views the watches shown by Hauptman in his article on the "The American Watch Company," one sees that the universal dials could have been used also on 10 size P. S. Bartlett, 10 size Appleton, Tracy, 16 size Appleton, Tracy, 14 size Crescent Garden, 14 size Export, 8 size Riverside and 8 size William Ellery models. (15)

In 1884, the Waltham dial room was under the direction of Charles Moore, who was first employed by the company in 1859. The Waltham dials of 1884 had a copper foundation, to which the feet were brazed. Pulverized enamel was then laid on the dial and it was "fired." From there, the dials were painted. The painters used camel hair brushes to apply enamel, and machines applied the minute marks. The painting function involved several operations and three firings of the dial. (16)

FIGURE 55   *Enameling and Firing Watch Dials, 1870*

Enter Daniel O'Hara

A technological improvement changed the course of dial making in the 1880's.   The Swiss developed a process (intaglio) for taking a print from a still die and transferring it to the dial.   Daniel O'Hara, the foreman of the Waltham case department must have learned of this process.   He abandoned his 10 years of experience in making cases to try to manufacture dials using the new technique.

Daniel O'Hara was born in Newport, Kentucky. He first worked as a casemaker at the Duhme Jewelry Company across the Ohio River in Cincinnati.   Next, he moved to the Dueber Case Company, and finally, in 1881, to Waltham to become foreman of its case department.   O'Hara obtained several patents for cases and parts, including the patent for a screwback watch case.   Early in 1890, the Waltham case department was closed.   O'Hara looked around for another job, and so it was that he linked up with Edwin D. Wetherbee--a dial maker.

Wetherbee was born in Hanover, New Hampshire in 1857. He moved to Waltham about 1878, when he began to make dials in a shop on Crescent Street. When O'Hara's shop was closed by the Waltham company, he joined Wetherbee in his shop. He was influenced favorably by O'Hara's "venturesome sprit and managerial ability." (17) They started what was first named the Waltham Watch Dial Company and later the Waltham Dial Company. As soon as they linked up, they began to experiment with the new Swiss process. They secured several of the Swiss machines, which they modified to allow the use of various colored enamels on one dial (remember that the Swiss machines could only handle two colors at one time.

Fortunately, they were making their technological mark very close to two sources of demand: the Waltham Watch Company and the United States Watch Company of Waltham. In 1891, they received orders from both companies. The next year, Wetherbee left the Waltham Dial Company to establish a similar department in the Trenton Watch Company factory. O'Hara then changed the name of his company to the O'Hara Dial Company. This name was changed later to the O'Hara Waltham Company and the O'Hara Waltham Manufacturing Company.

In the following year, O'Hara built a new factory on Rumford Avenue to take care of his growing business. It was 3 story, measuring 125 feet by 150 feet. Joseph O'Hara, (a brother ?) was made Superintendent and Daniel O'Hara, owner and manager. When the new factory opened, it employed 300 persons. They were required to satisfy a growing demand for dials for gas meters, counting meters, ash trays, bicycle name plates and for political campaign buttons. (18)

When Daniel O'Hara died in 1912, his company was producing few watch dials. In the following year, the O'Hara Waltham Dial Compamy was incorporated. Arthur Lyman was appointed President;

FIGURE 56 *A Fancy Waltham Dial. This Chronometro Victoria dial is attached to a AWW Co. watch, number 10,429,069 (1901). It is a 14 size, 15 jewel export Model 1897, with an Estrella 14K case (94367).*

Charles F. Stone, Treasurer and Clarence French, Secretary. Eliot O'Hara, Daniel's son, was made Superintendent.

Business rejuvenated during World War I, especially compass dials. But in 1927, Laurence M. Lombard, Dudley B. Wallace and Edward C. Clark obtained the use of the company name from Eliot O'Hara. The O'Hara estate sold out just before the crash of 1929. The company was sold again in 1933 to Henry T. Jackson, Francis J. Kelly and C. Russell Walton. They continued to make enamel products under Jackson's direction. (19)

Watch Hands

Prior to the American Civil War, watch hands were made by hand, and many of the parts were imported. Although Americans had the necessary skills for making hands, these could not be produced in sufficient quantities for the watch industry that was soon to grow rapidly. The new American Watch Company attempted to manufacture hands by machine, but it became discouraged at the technical obstacles to be overcome. The company next gave the job to one of its foremen, Bardwell A. Goodell. Goodell sought help from his brother-in-law, James. H. Winn, a machinist of the Sibley Machine Company in Waltham.

The result of this collaboration was the formation, in September 1868, of the firm of Goodell and Winn. Since they were both machinists and inventors, they started in a barn to make machinery to produce watch hands automatically. The first problem to overcome was the lack of power to run the machines. They moved to Cummingsville, near Woburn, Massachusetts, to take advantage of its water power, but when the flow was insufficient, they moved again to Winchester--their permanent location. They tore down an old wood mill and erected a two-story building, 20 feet by 38 feet. (20)

The company was a large success. Goodell and Winn was the first firm to devote itself exclusively to the automatic production of watch hands. When watch companies were established in the 1860's and after, they were content to import watch hands from Europe, but when the machine- made hands were produced at lower cost, the demand for Goodell and Winn hands propelled it into first place in the watch industry. But in 1876, the American Watch Company established its own watch hand department, thereby reducing that part of the Goodell and Winn business.

In 1896, Goodell sold his share in the business to Frank H. Winn, who was the second son of James H. Winn, the founder, and the company name was changed to J. H. Winn & Sons. Sales continued to expand, and in 1900, a new three-story factory was built on the site of the original wood building. This stone building was expanded in 1910 and 1920, but soon conditions in the watch industry worked against the company. Demand for their fine-quality "glossed" hands fell to foreign suppliers, who had much lower costs. A technological development came to the rescue: the application of electricity to the radio and automobile industries. Soon J. H. Winn & Sons was enjoying a large volume of business from making hands for clock radios and automobile clocks. In time, business expanded to the electronics industry.

FIGURE 57 *The Winn Factory, 1932*

References

1. Ernest A. Cramer, "The Watch Case and the Watch," NAWCC *Bulletin*, VII (Feb. 1957), p. 400.

2. Cooksey Shugart, *The Complete Guide to American Pocket Watches* (Cleveland: Overstreet, 1981), p. 65.

3. Warren H. Niebling, "A History of the American Watch Case," NAWCC *Bulletin*, IV (Feb. 1970), p. 168.

4. Cramer, "The Watch Case," p. 404.

5. Niebling, "A History of the American Watch Case," p. 28.

6. Charles K. Aked, "The First English Watch Manufactory," NAWCC *Bulletin*, XXVIII (Feb. 1986), p. 60.

7. Charles W. Moore, *Timing a Century--A History of the Waltham Watch Company* (Cambridge: Harvard University Press, 1945), p. 56.

8. *The New York Times*, May 26, 1903, p. 16.

9. See *The New York Times*, August 9, 1912, p. 5; June 25, 1923, p. 26; and May 25, 1930, II, p. 18.

10. Niebling, "A History of the American Watch Case," p. 28.

11. H. C. Hovey, "The American Watch Works," *Scientific American*, LI (Aug. 1884), p. 104.

12. Edwin B. Burt, " Waltham Dial Company," NAWCC *Bulletin*, VII (April 1956), p. 124.

13. W. Barclay Stephens, "The Adventure of the Lord's Prayer," NAWCC *Bulletin*, VI (Apr. 1954), 130-138.

14. Adin E. Mathews, "Answer Box," NAWCC *Bulletin*, XXII (Aug. 1980), p. 412.

15. Wesley R. Hauptman, "The American Watch Company," NAWCC *Bulletin*, XI (April 1964).

16. Hovey, "The American Watch Works," p. 104.

17. Burt, "Waltham Dial Company," p. 124.

18. Edmund L. Sanderson, *Waltham Industries* (Waltham: Waltham Historical Society, 1957), p. 113.

19. Still again the literature is confusing as to dates and personnel. Burt states that the O'Hara Watch business was sold in 1923 to Ralph A. Doe, who did business under the D & G Dial and Enameling Company name, and that the remainder of the company was sold in 1926 to R. L. Nimms. Obviously, this

history conflicts with that presented by Sanderson.

20. James W. Gibbs, "Watch Hands: The History of J. H. Winn Inc.," NAWCC *Bulletin*, XI (Apr. 1964), p. 224-227.

PART IV

End Matters

APPENDIX A

A Directory of Massachusetts Watchmakers

Abbott, Samuel, Boston, 1820
Adams, A.S., Boston, 1850
Adams, John P., Newburyport, 1835
Adams, Nathan, Danvers, 1755
Adams, Samuel, Boston, 1825
Adams, Thomas F., Boston, 1810
Adams, William, Boston, 1815
Alden, William, Fitchburg, 1885
Andrew, John, Salem, 1760
Allen, James, Boston, 1680
Allen S. D., Westfield, 1835
Almy, James, New Bedford, 1835
Appleton, George B., Salem, c1859-1865
Appleton, George B., Salem, 1860
Appleton, James, Marblehead, 1825
Archer, William, Jr., Salem, 1845
Ariettis & Co., Marblehead, 1870
Arthur, H. G., Boston, 1830
Ashby, James, Boston, 1769
Aspinwall, Zalmon, Boston, 1810
Aston, Thomas, Newburyport, c1860
Atkinson, James, Boston, 1755
Atwood, B. W., Plymouth, c1860
Austin, Josiah, Salem, c1835-1855
Austin, Josiah, Salem, 1840
Avery, John, Boston, 1726

--B--
Babcock, Alvin, Boston, 1810
Bacon Silas D., Boston, 1816-1893
Bacon & Smith, 4 Elm St., Boston,
    1845-1853
Badley, Thomas, Boston, 1710
Badlam, Stephen, Boston, 1770
Bagnall, Benjamin, Boston, 1710
Bagnall, Benjamin, Jr., Boston, 1740
Bagnall, Samuel, Boston, 1750
Bagnall, William, Boston, 1820
Bailey, Calvin, Hanover, 1795
Bailey, John I, Hanover, 1750
Bailey, John II, Hanover, 1790
Bailey, John III, Hingham, 1825
Bailey, Joseph, Hingham, 1820
Bailey, Lebbeus, Hanover, 1800
Baker, David, Newbedford, 1840-1850
Baker, George, Salem, 1790
Baker, Stephen, Salem, 1800
Baker, William, Boston, 1820
Balch, Benjamin, Salem, 1800
Balch, Chas. H., Newburyport, 1800
Balch, Daniel, Newburyport, 1760
Balch, Daniel, Jr., Newburypt, 1800
Balch, James, Salem, 1830

Balch, Moses P., Lowell, 1830
Balch, T. H., Newburyport, 1800
Baldwin, Jabez, Salem, 1795
Baldin, Jedidiah, Northampton, 1800
Barker, Jonathan, Worcester, 1800
Barker, William, Boston, 1810
Barnes, Horace, Boston, 1855
Barrell, Colborn, Boston, 1772
Barrett, Humphrey, Bolton, 1810-1885
Barton, John, Haverhill, c1840-1850
Barton, John, Salem, c1849
Barton, John, Salem, Newburyport, 1850
Barton, Joseph, Stockbridge, 1764
Barton, Wm. C., Salem, 1840
Bartow, John, Haverhill, 1850
Batchelder, Andrew, Danvers, 1810
Batchelder, Ezra, Jr., Danvers, 1825
Bates, James, Haverhill, c1860-1880
Bates, James C., Haverhill, 1885
Bateson, John, Boston, 1710
Baynes, B. B., Lowell, 1835
Beals, J. J., Boston, 1875
Beals, William, Boston, 1840
Beals, William, 21 Merrimac St.,
    Boston, c1853
Beath, John, Boston, 1805
Beatley, Ralph, Chelsea, c1840
Belknap, Ebenezer, Boston, 1810
Belknap, Wm. A., Boston, 1820
Bell, Samuel, Boston, 1813
Bemis, Samuel, Boston, 1820
Bemis, Samuel A., Boston, 1793-1881
Bentley, Thomas, Gloucester, 1790
Bichault, James, Boston, 1730
Bigelow, A. O., Boston, 1864
Bigelow & Bros., Boston, c1840
Billings, I., Acton, 1770
Billings, L., Northampton, 1837
Blaisdell, David, Amesbury, 1730
Blaisdell, David II, Amesbury, 1750
Boardman, J., Newburyport, 1850
Bochemsde, Frederick, Boston, 1815
Bodeley, Thomas, Boston, 1720
Bolkman, Ebenzer, Attleboro, 1740
Bond, Charles, Boston, 1835
Bond, William, Boston, 1800
Bottomley, T., Boston, 1784
Bowen, George A., Boston, 1840
Boyce, Benjamin M., Boston, 1880-1895
Boyce, Benjamin M., Boston, 1885
Boyd, John, Lowell, 1860
Boyd, John, Lowell, c1860
Boyter, Daniel, Boston, 1800

Brabner, William A., Boston, 1825
Brackett, J. R., Boston, 1842
Brackett, Jeffrey, 69 Washington St.,
    Boston, 1815-1876
Bradley & Barnes, Boston, c1850
Brand, James, Boston, 1711
Brand, John, Boston 1711
Bridgdon, C.H., Canton Junction, c1890
Brigden, C. H., Canton Jct., 1890
Briggs, Nathaniel, Boston, 1820
Brigham, L. S., Marlborough,
    c1832-1903
Brooks, William P. B., Boston, 1855
Brown Edw. L., Newburyport, 1860
Brown, Edward, Newburyport, c1860
Brown, Gawen, Boston, 1780
Brown, John James, Andover, 1840
Brown, John W., Newburyport, 1830
Brown, Nelson H., Boston 1900
Brown, Samuel A., Lowell, 1835
Browning, Samuel, Boston, 1815
Browning, Samuel, 77 Broad St., Boston
Buerk, J. E., Boston 1860
Buerk, J. E., Boston, c1860
Bullard, Charles, Boston, 1825
Bunker, Benjamin, Nantucket, 1825
Burkmar, Thomas, Boston, 1776
Burnap, Daniel, E. Windsor, Ct., 1800
Burr, Jonathan, Lexington, 1835
Burroughs, Thomas, Lowell, 1834
Burtt, Thomas F., Stoneham, 1855

    --C--
Carleton, Jas. H., Haverhill, 1853
Carleton, Michael, Bradford, 1800
Carpenter, C. H. Middleboro, 1870
Carpenter, C. H., Middleboro, c1860
Carpenter, Samuel, Flushing, N.Y.,
    1858
Carr, Frank B., FallRiver, c1860
Carr, James, Lowell, 1834
Chamberlain, Benjamin M., salem,
    c1840-1870
Chapman, C. H., Easthampton, 1860
Chapman, C. H., Easthampton, c1860
Chase, Samuel, Newburyport, c1850
Chase, Samuel, Newburyport, 1851
Chase, Wm. H., Salem, 1840
Chase, William H., Salem, c1840
Child, True W., Boston 1823
Chittenden, Austin, Lexington, 1834
Cito, J. C., Boston, 1820
Clapp, John, Roxbury, 1800
Clapp, Preserved, Amherst, 1765

Clark, Thomas, Boston, 1764
Clark, William H., Palmer, c1850-1860
Cogswell, Henry, Salem, 1845
Cogswell, John C., Salem, 1853
Cogswell, Robert, Haverhill, 1830
Collins, Elijah, Boston, 1727
Conant, Elias, Bridgewater, 1790
Conant, Nath. P., Danvers, 1850
Conant, Nathaniel Peabody, Danvers,
    c1850
Coning, Richard, Boston, 1796
Cook, Benj., Northampton, 1846
Cook, Benjamin E., Northampton,
    c1840-1850
Cranch, Richard, Salem, 1760
Crandell, J. R., Salem, 1875
Crane, Aaron D., Caldwell, N.J., 1830
Crane, John E., Lowell, 1850
Crane, Simeon, Canton, 1810
Crane, William, Canton, 1780
Crawford, William, Oakham, 1775
Crawley, Abraham, Boston, 1815
Crehore, Chas. C., Roxbury, 1815
Crosby, C. A., Boston, 1875
Crosby, Samuel T., Boston, 1856
Cross, Theodore, Boston, 1770
Cummens, William, Boston, 1800
Cummens, William Jr., Boston, 1810
Currier, Edmund, Salem, 1840
Currier, John, Salem, 1831
Currier, Eliphalet, Haverhill, 1800
Curtis, Benjamin B., Boston, 1810
Curtis, Lemuel, Concord, 1820
Curtis, Samuel, Boston, 1820
Curtis, Thomas, Lowell, c1860
Curtis, W., Newburyport, 1825
Cushing, Theodore, Hingham, 1808
Cutler, Amos, Boston, 1803-1894

    --D--
Dakin, James, Boston, 1796
Dalrymple, James, Salem, 1795
Davenport, James, Salem, 1784
Davis, Ari, Boston, 1840
Davis, Ari, Boston, c1840
Davis, D. P., Boston, 1845
Davis, David Porter, Roxbury, 1840-185
Davis, John W., Salem, 1824
Davis, Samuel, Boston, 1830
Dawes, Robert, Boston 1842
Dean, George, Salem, 1800
Dean, William, Salem, 1800

DeGiy, Lewis, Boston, 1800
Delfin, Joseph, Boston, 1871
Delphin, Joseph J., School St.,
    Boston, c1871
Delvin, Mark H., Salem, 1864
Delvin, Mark H., Salem, c1864
Derby, Charles, Salem, 1850
Derby, Charles, Jr., Salem, c1870
Derby, Charles, Jr., Salem, 1859
Deverell, John, Boston, 1790
Dexter, Dana, Boston, 1870
Dexter, Dana, Roxbury, c1860
Doanne, John, Scituate, 1790
Dodge, George Jr., Salem, 1837
Dogget, John, Boston, 1810
Dogget, Samuel, Boston, 1810
Dole, H. L. & Co., 4 Merrimack St.,
    Haverhill, c1865-1877
Doull, James, Charlestown, 1800
Drown, Charles L., Newburyport, 1850
Drown, Charles L., Newburyport, c1850
Dudley, Thomas, Boston, 1807
Dumesnil, Anthony, Boston, 1800
Durand, Asher B., Boston, 1840
Dutch, Stephen, Boston, 1800
Dutch, Stephen, Boston, 1810
Dyar, George W., Boston, 1825
Dyar, Harrison, G., Concord, 1820
Dyar, Joseph, Concord, 1810
Dyar, Warren, Lowell, 1831

--E--

Eastman, Joseph, Boston, 1880
Eaton, James, Boston, 1807
Eaton, John H., Boston, 1809
Eaton, Leander, Barre, 1822-1901
Eaton, Samuel A., Boston 1825
Edger, Sylvester, Roxbury, 1830
Edson, Jonah, Bridgewater, 1815
Edwards, Abraham, Ashby, 1794
Edwards, John, Ashby, 1809
Edwards, Nathan, Acton, 1825
Edwards, Samuel, Sr., Ashby, 1760
Edwards, Samuel, Jr., Ashby, 1806
Elliott, Hazen, Lowell, 1832
Emery, Samuel, Salem, 1809
Emery, Samuel, Salem, 1810-1855
Emmet, Edward T., Boston, 1764
Emmons, C. G., Boston, 1842
Emmons, C. G., Boston, c1844
Entwistle, Edmund, Boston, 1742
Essey, Joseph, Boston, 1712

--F--

Fahrenbach, Pius, Boston, 1850
Fahrenbach, Pius, Boston, C1850
Fairbanks, James H., Fitchburg, 1878
Fairbanks, Joseph O., Newburyport,
    1860
Fairman, Gideon, Newburyport, 1800
Fales, G. S., New Bedford, 1820
Fales, James, New Bedford, 1815
Fales, James, Jr., New Bedford, 1830
Fales, James, & Giles, New Bedford,
    c1840
Farmer, M. G., Salem, 1849
Farmer, M. G., Salem, c1849
Farnham, Henry, Boston, 1780
Farnham, Rufus, Boston, 1780
Fellows, Ignatius W., Lowell, 1834
Fellows, James K., Lowell, 1832
Felton, A. C., Boston, 1860
Fenno, James, Lowell, 1834
Fisk, Samuel, Boston, 1790
Fisk, William, Boston Neck, 1800
Fitch, Jonas, Pepperell, 1780
Flagg, Seth, Springfield, 1840
Flagg, Seth, Springfield, c1840
Fole, Nat., Northampton, 1819
Folger, Peter, Nantucket, 1650
Folger, W., Jr., Nantucket, 1810
Follet, Marvill M., Lowell, 1835
Foster, George, Boston, c1840
Foster, George B., Boston, 1842
Foster, Nathaniel, Newburyport, 1840
Foster, Thomas, Newburyport, 1840
Foster, Thomas Wells, Newburypt, 1860
Fowle, John, Boston, 1810
Fowle, J. H., Northampton, 1850
Fowle, Nathaniel, Boston, 1803
Fowle, William B., Aubumdale, c1879
Fowler, John C., Boston, 1842
Fox, Philetus, Boston, 1842
Frary, Obediah, Southampton, 1760
French, Lemuel, Boston, 1800
Frost, Benjamin, Reading, 1850
Frost, Jesse, Lynn, 1835
Frost, Jonathan, Reading, 1840
Frye, James, Haverhill, 1853
Fuller, Artemas, Lowell, 1845
    c1840-1850
Furnival, James, Marblehead, 1780

--G--

Gale, James, Salem, 1815
Galiay, John P., Boston, 1825

Garrish, D. D., Boston 1854
Gates, Zacheus, Harvard, 1820
Geddes, Charles, Boston, 1776
Gere, Issac, Northampton, 1795
Gibbons, Thomas, Boston, 1739
Gibbs, Benj., Newburyport, 1820
Gill, Caleb, Hingham, 1785
Gill, Leavitt, Hingham, 1790
Glover, Wm., Boston, 1820
Goddard, Benj., Worcester, 1849
Goddard, D. & Co., Worcester, c1850
Goddard, Dan., Shrewsbury, 1810
Goddard, George S., Boston, 1820
Goddard, L., Shrewshbury, 1810
Goddard, Nicholas, Northampton, 1795
Goddard, Parley, Wrcester, 1825
Goodfellow, John, Boston, 1800
Gooding, Alanson, New Bedford, 1820
Gooding, Henry, Boston, 1820
Gooding, John, Plymouth, 1825
Gooding, Joseph, Dighton, 1830
Goodnow, S. B. Fitchburg, 1840
Goodwin, Henry, Boston, 1820
Goodwin, Wallace, No. Attleboro, 1840
Gordon, Smyley, Lowell, 1832
Goron, Thomas, Boston, 1759
Grant, William, Boston, 1820
Gray & Basil, Boston, c1860
Green, John, Boston, 1796
Green, Samuel, Boston, 1820
Green, Samuel Jr., Boston, 1825
Greenough, N.C., Newburyport, 1870
Griswold, H., Leominster, 1800
Grubb, Wm. Jr., Boston, 1816

--H--
Hale, Joshua, Lowell, 1840
Hale, William C., Salem, 1850
Hall, Asa, Boston, 1806
Hamilton, James, Boston
Hampton, Samuel, Chelsea, 1840
Hampton, Samuel, Chelsea, c1840
Haneye, Nathaniel, Bridgewater, 1790
Hannum, John, Northampton, 1837
Harding, Newell, Haverhill, 1850
Harding, Newell, Haverhill and Boston,
   c1840
Harmson, Henry, Marblehead, 1720
Harrington, Henry, Salem, 1855
Harrington, Henry, Salem, c1855
Harrington, Sam., Amherst, 1842
Harris & Stanwood, Boston, C1842
Hartwell, A., Worcester, 1872

Harwood Bros., Boston, c1865
Hastings, T. D., Boston, 1854
Hastings, T. D., Boston, c1854
Hatch, George D., No. Attleboro, 1865
Hatch, John B., Attleboro, 1880
Hayden Samuel, Boston, 1786
Healy, John W., Worcester, 1850
Heffards, S. M., Middleboro, 1850
Hewitt, A. E., No. Bridgewater, 1860
Hill, John B., Beverly, 1850
Hill, John B., Beverly, c1850-1860
Hill, Wm. F., Boston, 1810
Hiller, Joseph, Boston, 1770
Hills, Charles C., Haverhill, 1853
Hobart, Aaron, Abington, 1770
Hobbs, Nathan, Boston, 1842
Holbrook, H., Medway, 1830
Holden J., Boston, 1830
Hollister, J. H., Greenfield,
   c1850-1860
Holmes, Aaron, Boston, 1842
Holway, Philip, Falmouth, 1850
Homan, Samuel, Marblehead, 1850
Homan Watch Co., Boston, c1890
Horn, E. B, Boston, c1840
Horn, Eliphalet, Lowell, 1835
Hovey, Cyrus, Lowell, c1860
Hovey, Cyrus, Lowell, 1870
Howard, Albert, Boston, 1878
Howard, Albert, Boston, 1881-1893
Howard, Edward, Boston, 1842
Howard, J., Boston, c1870
Howard, J., Boston, 1875
Howard, A. & Co., Boston, 1878
Howard, Wm., Boston, 1815
Howe, Elias, Boston, c1843
Howe, Jubal, Boston, 1830
Hubbard, C., Boston, 1845
Hubbard, Daniel, Medford, 1820
Huntington & Church, Springfield, 1800
Huntress, L., Lowell, 1840
Hutchins, Abel, Roxbury, 1786

--I--
Ingersoll, Daniel G., Boston, 1805
Ingersoll, Daniel III, Boston, 1800

--J--
Jackson, John, Boston, 1791
Jackson, Thomas, Boston, 1770
James, Joshua, Boston, 1815
Jarvis, John J., Boston, 1787
Jenkins, Osmore, Lowell, 1840

Jepson, Wm., Boston, 1830
Johnson, Andrew, Boston, 1805
Johnson, Caleb, Boston, 1805
Johnson, Eli, Boston, 1821
Johnson, Thomas S., Boston, 1862
Jones, Albert, Greenfield, 1850
Jones, Ezekiel, Boston, 1790
Jones, J. B., Boston, 1822
Joseph, Isaac, Boston, 1823

--K--
Kehew, Wm. H., Salem, 1855
Keith, Wm. H., Shrewsbury, 1810
Kelley, Allen Sandwich, 1820
Kelly, Ezra, New Bedford, 1860
Kelly, John, New Bedford, 1836
Kemlo, Francis, Chelsea, 1840
Kendall, D. C., Boston, 1842
Kenney, Asa, West Milbury, 1800
Kettel, J.V., Boston, 1821-1896
Keyes, Rufus, Lowell, 1833
Kilburn, Hiram, Lowell, 1837
Kimball & Goald, Haverhill, 1865
Kimball, John Jr., Boston, 1820
Kimball N., Boston, 1820
King, W. B., Boston, 1810
Kirkham, James, Springfield, 1821-1893
Kirkland, Sam. W., Northampton, 1840
Knapp, Jesse, Boston, 1830
Knower, Daniel, Roxbury, 1800
Kramer M. & Co., Boston, 1840

--L--
Lakeman, Ebenezer K., Salem, 1820
Lamb, Cyrus, Oxford, 1830
Lamb, T.M., Worcester, c1855
Lamson, Charles, Salem, c1850
Lamson, Chas., Salem, 1845
Lane, D.H., Gloncester, c1870
Langdon, William G., Boston, 1862
Lanny, David F., Boston, 1789
Larkin, Joseph, Boston, 1845
Larkin, Joseph, 5 Endicott St.,
    Boston, c1840
Lawrence, George, Lowell, 1832
Leach, Caleb, Plymouth, 1780
Learned, Wm. B., Boston 1880
Learned, William, Boston, c1898
Leavitt, Dr. Josiah, Hingham, 1772
Leland, Caleb, Templeton, c1870
Leonard, Chas., Newburyport, 1850
Lloyd, Wm., Springfield, 1802
Locke, Edward, Boston, c1880

Lockwood, Wm., Charlestown, 1790
Lombard, Daniel Jr., Boston, 1830
Loring, Henry W., Boston, 1820
Loring, Joseph, Sterling, 1800
Lovett, James, Mendon, 1775
Lovett, Wm., Boston, 1850
Lovis, Capt. Jos., Hingham, 1790
Low, Daniel, Boston, 1859
Low, Daniel, 207 Essex St., Boston,
    c1845
Luscomb, Samuel, Salem, 1770
Lyman, C. A., Palmer, 1880
Lyman, Roland, Lowell, 1835

--M--
M'Clure, David, Boston, 1810
M'Clure, John, Boston, 1825
MacFarlane, John, Boston, 1796
McHugh, James, Lowell, 1860
McKay, H., Haverhill, 1905
MacKay, Crafts, Boston, 1789
MacKay, H., Haverhill, 1925
Macomber, A.J., Newton Center,
    1826-1895
M'Pherson, Sweeney, E. Boston, 1830
Malloy, George, Lowell, 1859
Manning, Richard, Ipswich, 1750
Manning, Thomas, Salem, 1690
Marsh, Oliver, Roxbury, c1849
Martin, Valentine, Boston, 1842
Mason, H. G., Boston, 1845
Mason, Timothy B., Boston, 1830
Maxwell, James, Boston, 1730
Meacham, W.J., Holyoke, c1850
Metcalf, F., Hopkinton, 1825
Metcalf, Luther, Medway, 1800
Minot, J., Boston, 1805
Mitchell, Phineas, Boston, 1815
Mitchelson, David, Boston, 1774
Molloy, George, Lowell, 1850
Monroe, John, Barnstable, 1860
Moore, George H., Lynn, 1860
Morgan, Luther S., Salem, 1845
Morgan, Theodore, Salem, 1835
Morris, Elijah, Canton, 1820
Morris, Henry, Canton, 1820
Morris, Wm., Grafton, 1770
Morse, Joseph, Walpole, 1850
Morse, Moses, L., Boston, 1813
Moseley, Charles, Boston, 1860
Moseley, Charles S., Waltham, 1852
Moseley, Robt. E., Newburyport, 1848
Moulton, Joseph, Newburyport, 1872

Moulton, Francis, Lowell, 1930
Moulton, Francis E., Lowell, 1832
Moulton, Jos., Newbury, 1730
Mozart Don J., Boston, 1823
Mulliken, Benj. Bradford, 1745
Mulliken, John, Bradford, 1730
Mulliken, Jonathan, Bradfd., 1830
Mulliken, Jonathan, Newburyport, 1772
Mulliken, Jos., Newburyport, 1804
Mulliken, Jos., Salem, 1795
Mulliken, Jos. Concord, 1777
Mulliken, Nath. I, Lexington, 1767
Mulliken, Nath. II, Lexington, 1770
Mulliken, Nath., Newburyport, 1776
Mulliken, Nathaniel, Lexington, 1820
Mulliken, Robt., Bradford, 1640
Mulliken, Sam. I, Bradford, 1740
Mulliken, Sam. II, Salem, 1790
Mulliken, Sam. Haverhill, 1850
Mulliken, Sam., Newburyport, 1770
Munroe, Daniel, Concord, 1796
Munroe, Deacon Nehemiah, Boston, 1790
Munroe, Nath., Concord, 1810
Munroe, Wm., Concord, 1800
Munyan, A. H., N'thampton, 1848
Munyan, A. H., Northampton, c1845

--N--

Neilson, George, Boston, 1830
Nelson, John A., Boston, 1825
Newall, Thos., Sheffield, 1809
Newcomb, Henry, Lowell, 1837
Newcomb, Thomas, Boston, 1810
Newell, A., Boston, 1785
Newell, Thos., Sheffield, 1815
Newhall, Fred. A., Salem, 1853
Newhall, Frederick A., Salem, c1850
Newhall, William, Boston, c1850
Newman, John, Boston, 1764
Newton Isaac L., Salem, 1796
Nicholas, Wm C., Winchendon, 1840
Noble, Philander, Pittsfield, 1830
Nolen, Spencer, Boston, 1806
Norton, Samuel, Hingham, 1785
Noyes, L. W., Boston, 1841
Nye, Wm. F., New Bedford, 1862

--O--

Oakes, Tila, Ashby, 1800
O'Connell, Maurice, Boston, 1842
Orne, R. S., Boston, 1840
Orne, R. S., Boston, c1840

Osborne, John, Lynn, 1827
Osgood, John Jr., Boston, 1825
Osgood, Orlando F., Haverhill, 1845

--P--

Packard, Isaac, Brockton, 1825
Palmer, Bacheller & Co., 394
Washington
    St., Boston, c1875
Palmer, D. D., Waltham, 1870
Palmer, D. D., Waltham, 1864-1875
Palmer, Dolphus D., Waltham, 1839-1907
Palmer, Sam., Dedham, 1800
Parker, Daniel, Boston, 1760
Parker, Isaac, Deerfield, 1780
Peaseley, Robert, Boston, 1735
Peck, Elijah, Boston, 1789
Peck, Moses, Boston, 1753
Peele, J. B., Salem, 1818
Penniman, John R., Boston, 1807
Perrigo, James, Wrentham, 1760
Perrigo, Jas. Jr., Wrentham, 1800
Perry, Albert, Salem, 1850
Perry, Albert, 25, Carlton St., Salem,
    c1850
Pettee, Simon, Wrentham, 1840
Pickett, Rich. Jr., Newburyport, 1836
Pitman, Wm. R., New Bedford, 1836
Platt, Samuel, Boston, 1825
Pomeroy, Eltweed, Bolton, 1650
Pond & Barnes, Boston, c1845
Pond L. A., Boston, 1845
Pond, Wm., Boston, 1842
Pope, Joseph, Boston, 1790
Pope, Robert, Boston, 1786
Potter, Albert E., Boston, 1870
Porter, George, Boston, 1835
Potter, John, Brookfield, 1785
Pratt, Daniel Jr., Reading, 1839
Pratt, Wm., Boston, 1842
Proctor, G. K., Beverly, 1860
Pulsifer, F., Boston, c1850
Pulsifer, F. L., Boston, 1854

--R--

Rand, Daniel, Boston, 1830
Rawson, Jason R., Holden, 1839
Ray, Daniel, Sudbury, 1790
Raymond, Benj., W. Elgin, Ill., 1850
Rea, Archelaus, Jr., Salem, 1791
Reading, Joseph, Boston, 1855
Reed, Ezekiel, Brockton, 1790°
Reed, George P., Boston, 1865

Reed, George P., Boston and Melrose,
    1828-1901
Reed, Simeon, Cummington, 1770
Regally, M., Boston, 1842
Rice, Luther G., Lowell, 1835
Rice, Oran S., Fitchburg, 1878
Rice, Phineas, Boston, 1830
Rich, Wm., Salem, 1850
Richards, Frank L., Boston, 1850
Richardson, John, Boston, 1805
Robbins, Royal E., Waltham, 1858
Roberts, Joseph, Newburyport, 1850
Robinson, J. A., Lowell, 1835
Robinson, Obed., Attleboro, 1790
Rogers, Caleb, Newton, 1795
Rogers, Isaac, Marshfield, 1815
Rogers, John, Billerica, 1765
Rogers, Wm., Boston, 1860
Roulstone, John, Boston, 1770
Roydor, Francis, Boston, 1813
Russell, A. L., Lynn, 1845
Russell, Charles H., Lynn, 1848
Russell, D. N., Greenfield, 1840
Russell, Maj. John, Deerfield, 1765
Russell, Sam., Middletown, 1846
Russell, Samuel, Middletown, c1846

--S--

Sample, Wm., Boston, 1850
Sample, William, Boston, C1850
Sargeant, Ebenezer, Newburypt, 1790
Sargeant, Thos., Springield, 1850
Sargent, Joseph, Springfield, 1825
Sawin, John, Boston, 1825
Sawyer, Joel, Bolton, 1805-1897
Sawyer, Levertt, Salem, 1830
Sawyer, Syvanus, Fitchbury, c1875
Schollet, John B., Boston, 1796
Seward, Joshua, Boston, 1840
Shaw, David, Plainfield, 1835
Shearman, Martial, Andover, 1835
Shearman, Martin, Hingham, 1821
Shepherd, Nath., N. Bedford, 1810
Shepherd, Nathaniel, New Bedford, 1872
Sherwin, Wm., Buckland, 1830
Sibley, Gibbs, Sutton, 1786
Sibley, Col. Timothy, Sutton, 1780
Sibley, Richard S., Boston, 1845
Sibley, Stephen, G. Barrington, 1790
Silver, Daniel, Lawrence, 1860
Simnet, John, Boston, 1769
Skinner, Alvah, Boston, 1830
Smith, A., Boston, 1854

Smith, Daniel T., Salem, 1880
Smith, Daniel T., Salem, 1850
Smith, Hiram, Boston, 1852
Smith, Hiram, 32 Washington St.,
    Boston, c1859
Smith, Jesse Jr., Salem, 1810
Smith, Jesse R., Salem, 1857
Snatt, John, Ashford, 1710
Snow, Jeremiah, Springfield, 1820
South, James, Charlestown, 1810
Spence, John, Boston, 1825
Stanley, Phineas, Lowell, 1837
Stanwood, H.B., Boston, 1856
Stennes, E. O., Weymouth, 1940
Stevens, B. F., Peabody, 1845
Stevens, George M., Boston, 1875
Stevens, John, Salem, 1825
Stevens, S., Lowell, 1853
Stevens, S., Lowell, c1853
Stewart, Daniel, Salem, 1780
Stickney, Moses, Boston, 1823
Stiles, Samuel, Northampton, 1785
Stone, Ezra, Boston, 1810
Stone, Jasper, Charlestown, 1845
Storrs, Nath., Northampton, 1791
Storrs, Thomas, Boston, 1803
Stowe, L. G., So. Gardner, 1850
Stowell, A., Charlestown, 1850
Stowell, Abel, Worcester, 1800
Stowell, Abel Jr., Boston, 1830
Stowell, John, Charlestown, 1830
Stowell, John, Boston, 1820
Stowell, John J., Boston, 1831
Stratton, Chas., Worcester, 1835
Stratton, N. P., Waltham, 1850
Streeter, Gilbert L., Salem, 1846
Studley, David, Hanover, 1806
Studley, Luther, No. Bridgewater, 1840
Sumner, Wm., Boston, 1685
Sutton, Enoch, Boston, 1830
Swan, Benjamin, Haverhill, 1810
Sweet, James S., Boston, 1842
Sykes, Wm., Beverly, 1825

--T--

Taber, Elnathan, Roxbury, 1800
Taber, H., Boston, 1850
Taber, Stephen, N., New Bedford, 1798
Taber, Thomas, Roxbury, 1840
Tabor, L. A., Holyoke, 1860
Talbot, Sylvester, Dedham, 1815
Tappan, Ebenezer, Manchester, 1800
Tappan, Israel, Manchester, 1825

Tappan, J. F., Manchester, 1825
Tarbell, Edmund, Boston, 1842
Taylor, Richard, Boston, 1660
Terry, Thomas, Boston, 1815
Thatcher, Geo., Lowell, 1859
Thayer, Eli, Boston, 1810
Thayer, Eliphalet, Williamsburg, 1832
Thomas, Isaac Jr., Worcester, 1800
Thomequex, Peter, Northampton, 1802
Thompson, S. N., Roxbury, 1853
Thompson, S & N, Roxbury, c1850
Tifft, Horace, No. Attleboro, 1820
Timson, Wm. W., Newburyport 1860
Tisdale, E. D., Taunton, 1870
Titcomb, Enoch J., Boston, 1834
Tolman, Jeremiah, Boston, 1810
Townsend, David, Boston, 1800
Townsend, H., Conway, 1870
Townsend, Isaac, Boston, 1790
Trenchard, Richard, Salem, 1800
Trott, Andrew C., Boston, 1805
Trott, Peter, Boston, 1800
Turell, Andrew C., Boston, 1805
Turrell, Samuel, Boston, 1789

--V--
Valentine, Wm. B., Boston, 1821
Vanderwarker, C. H., Fitchburg, 1878

--W--
Wade, Chas., Boston, 1830
Walker, Isaac, Long Plain, 1800
Wall, W. A., New Bedford, 1825
Walley, John, Boston, 1800
Warren, Henry E., Ashland, 1915
Watson, G., Chelsea, c1845
Watson, J., Chelsea, 1840
Watson, John, Boston, 1842
Watson, John, Boston, c1842
Watts, Stuart, Boston, 1740
Webb, Isaac, Boston, 1710
Wells, Alfred, Boston, 1804
Welsh, Bela, Northampton, 1808
Wentworth, J. L., Lowell, 1835
Whidden, S. F., Boston, 1858
White, L. W., North Adams, 1850
Whiting, Samuel, Concord, 1810
Whitman, Ezra, Bridgewater, 1810
Whitney, Geo., Boston, 1803
Whitney, Moses, Boston, 1830
Whittemore, J., Boston, 1856
Wieland, Johann G., Salem, 1780
Wilbur, L. N., Fitchburg, 1878

Wilder, Ezra, Hingham, 1825
Wilder, Joshua, Hingham, 1810
Willard, Aaron, Grafton, 1770
Willard, Benj., Grafton, 1790
Willard, Benj. F., Boston, 1835
Willard, Ephraim, Grafton, 1760
Willard, Henry, Boston, 1825
Willard, John, Boston, 1804
Willard, John Mears, Boston, 1840
Willard, Simon, Grafton, 1775
Willard, Simon Jr., Roxbury, 1825
Willard, Zabadiel, Roxbury, 1841
William C., Salem, 1850
Williams, Hinds P., Boston, 1860
Williams, Nathaniel, Taunton, 1780
Williams, William, Boston, 1765
Winslow, Ezra, Westborough, 1860
Winslow, Jonathan, Harwich, 1800
Wood, B. B., Boston, 1841
Wood, David, Newburyport, 1810
Wood, Josiah, New Bedford, 1825
Wood, N. G., Boston, 1856
Woodworth, E. C., Boston, 1850
Wruck, F. A., Salem, 1864
Wyman, Benjamin, Boston, 1845

--Y--
Yeakle, Solomon, Northampton, 1830
Yeomans, Elijah, Hadley, 1771

Footnotes:

*
Some persons may have been clockmakers only. There is no way to
separate them from the watchmakers.

**
Includes only persons who were
active after 1840, when the MA watch
industry was in its infant stage.

Source: John Edwards, <u>The Complete
Checklist of American Clock and Watch
Makers</u> (Stratford, 1977);
<u>Massachusetts Register</u>; <u>Boston
Business Directory</u>.

U.S. WATCH PATENTS
MASSACHASETTS RESIDENTS

Inventor	Residence	Issue Date	Patent No.
Adams, J.P.	Ipswich	Jun. 24, 1873	140,231
Bigelow, J.K.	Springfield	Feb. 8, 1859	169,512
	Waltham	Nov. 2, 1875	22,914
Bingham, B.D.	Boston	Feb. 4, 1868	74,041
Briggs, L.C.	Boston	Jul. 30, 1878	206,533
Buck, D.A.A.	Worcester	May 21, 1878	203,999
			203,998
Buckland, E.H.	Springfield	May 5, 1868	77,579
			77,580
Champit, G.W.	Attleborough	Nov. 7, 1871	120,623
Church, D.H.	Waltham	May 11, 1875	163,161
	Waltham	July 3, 1883	280,719
	Waltham	Mar. 18, 1884	295,484
	Waltham	Feb. 17, 1885	312,253
	Newton	Aug. 4, 1885	10,631
	Boston	Oct. 13, 1885	328,289
	Newton	Apr. 6, 1886	339,378
	Newton	Aug. 24, 1886	347,994
	Waltham	Feb. 15, 1887	357,906
	Newton	Oct. 4, 1887	370,929
	Newton	Jan. 15, 1889	396,267
Cutter, D.G.	Waltham	Oct. 18, 1870	108,332
Cutter, F.R.	Somerville	May 4, 1886	341,095
Dennison, A.L.	Waltham	Jan. 11, 1859	22,550
		Dec. 11, 1860	30,873
		Jan. 1, 1861	31,009
Dill, T.B.	Boston	Aug. 23, 1870	106,561
Drinkwater, O.A.	Boston	Jun. 24, 1886	344,727

Eaton, L.	Worcester	Dec. 28, 1880	235,940
Ebelhave, W.H.	Waltham	Sept 27, 1910	971,496
Ebelhave, W.H. & J.A. Freund	Waltham	June 4, 1912	42,574
Eldridge, D.W.	Boston	Feb. 22, 1870	100,130
Elson, J.	Boston	Mar. 8, 1870	100,511
		Apr. 4, 1871	113,281
Fitch, E.C.	Newton	Aug. 26, 1884	304,264
		Apr. 14, 1885	315,755
		Apr. 14, 1885	315,756
		Apr. 28, 1885	316,767
		Jun. 9, 1885	319,691
		Aug. 18, 1885	324,675
		Jul. 5, 1887	366,085
		Dec. 24, 1889	417,999
		Mar. 25, 1890	424,191
Fitts, D.B.	Holliston	Nov. 30, 1858	22,174
Fogg, C.W.	Waltham	Feb. 2, 1864	41,461
		Feb. 14, 1865	46,343
Folsom, R.	Boston	Jun. 16, 1874	152,097
Freeland, A.F. & Whitten, C.H.	Waltham	Aug. 25, 1885	325,204
Freund, J.A.	Waltham	Apr. 25, 1905	788,399
		June 11, 1912	1,029,116
		May 17, 1921	1,378,408
		Sept 27, 1921	1,392,006
			1,392,007
Gardner, J., Jr.	Boston	Nov. 23, 1869	97,186
Gibson, C.D.P.	Boston	Feb. 1, 1870	99,428
Goldthwait	Long Meadow	May 6, 1890	427,072
Haldy, H.L	Waltham	Feb. 19, 1889	398,251
Hall, E.J.	Waltham	Nov. 13, 1883	288,233
		Sep. 30, 1890	437,345
Harwood, J.	Somerville	May 19, 1874	151,021

Hastings, G.	Waltham	May 26, 1860	52,849
		Feb. 27, 1866	65,097
		May 22, 1866	59,898
Hopkins, C.	Waltham	Aug. 13, 1875	167,174
Hopkins, J.R.	Waltham	July 20, 1875	161,513
Horn, E.B.	Boston	Jun. 11, 1867	65,747
		Dec. 24, 1867	72,641
Howard, E.	Boston	May 21, 1867	74,090
	Boston	Feb. 4, 1868	75,018
	Boston	May 31, 1868	64,874
Hunt, G.	Springfield	Oct. 10, 1871	119,768
Jewett, A.	Boston	Apr. 2, 1867	63,525
	Lynn	Aug. 24, 1875	166,990
Jones, E.A.	Boston	Feb. 2, 1869	86,411
Keller, A.L.	Springfield	Aug. 7, 1886	387,306
		Dec. 7, 1886	353,749
Lison, J.	Boston	Jun. 1, 1869	90,647
Logan, J.	Waltham	Nov. 10, 1885	329,915
		Nov. 10, 1885	329,916
		Jun. 1, 1886	342,917
	Boston	Jul. 19, 1870	105,467
		Nov. 6, 1888	392,442
		Jul. 9, 1889	406,655
Logan, J. & Church, D.H.	Waltham & Newton		
Manchester, E.L. & Bollen, J.A.	Springfield	Oct. 27, 1874	156,227
Margot, E.F.	Boston	Mar. 15, 1887	359,397
Marsh, E.A.	Newton	Feb. 8, 1887	357,398
		Mar. 29, 1887	360,234
		Sept 13, 1887	368,866
Mehl, W.B.	Waltham	May 5, 1908	886,387
		Oct. 6, 1908	900,183
Metzger, J.	Cambridge	Oct. 12, 1864	44,734

Morrill, C.F.	Boston	Aug. 10, 1886	347,252
		Feb. 22, 1887	358,403
		Mar. 29, 1887	360,105
		Sept 13, 1887	369,871
		Nov. 1, 1887	372,558
		Dec. 24, 1889	418,047
		Nov. 25, 1890	441,322
			441,434
			441,435
			441,436
Morrill, C.F. & Percival, D.C.	Boston	Jun. 12, 1888	384,380
Norcross, A.C.	Boston	May 31, 1881	242,358
O'Hara D.	Boston	Apr. 22, 1884	297,533
	Waltham	May 13, 1884	298,616
	Waltham		298,396
O'Hara, D.	Boston		298,615
		Aug. 19, 1884	303,881
		Dec. 9, 1884	309,158
		Dec. 16, 1884	309,354
		Mar. 3, 1885	313,360
		Apr. 28, 1885	316,814
		Jun. 2, 1885	10,605
		Apr. 27, 1886	340,814
		Dec. 7, 1886	353,706
			353,961
Olney, W.D.	Waltham	Jun. 10, 1890	429,699
Palmer, D.D.	Waltham	Jan. 8, 1889	395,754
Parker, F.	Weston	Jun. 7, 1887	364,370
Peabody, R.L.	Waltham	Jul. 17, 1883	281,295
Perry, E.H.	Boston	May 30, 1871	115,351
Plymtom, N.A.	Northborough	Mar. 12, 1867	62,775
Proctor, T.F.	Waltham	Jul. 11, 1882	260,783
		Jan. 6, 1885	310,313
Reed, G.P	Roxbury	May 27, 1802	35,389
	Waltham	July 1, 1856	15,251
	Walthan	Apr. 14, 1857	17,055
	Roxbury	Oct. 2, 1860	30,247

Reed, G.P	Roxbury	Apr. 9, 1861	21,999
	Roxbury	May 12, 1863	38,502
	Roxbury	Apr. 1, 1865	49,154
	Roxbury	Aug. 1, 1865	49,154
	Boston	Jun. 10, 1873	139,735
		May 26, 1874	151,243
	Melrose	Feb. 3, 1885	311,609
	Boston	Aug. 18, 1868	81,107
Rice, C.W. & Burbank, A.L.	Worcester	May 25, 1875	163,694
Rice, O.P. & Gerry, J.H.	Springfield	Jun. 9, 1868	78,693
Riggs, H.A.	Chelsea	Mar. 9, 1880	225,421
Rivett, E.	Boston	Feb. 3, 1880	224,227
Robbins, A.F.	Orange	Nov. 18, 1884	308,096
Rosenburg, B.	Waltham	Oct. 15, 1889	412,796
Schlegel, E. & W.	Boston	Feb. 6, 1877	187,180
Simmons, R.T.	Attleborough	May 13, 1879	215,250
St. John, G.B.	Boston	Jun. 2, 1885	319,145
Stratton, N.P.	Waltham	Jan. 3, 1860	26,715
Taft, R.L.	Waltham	Aug. 1, 1905	796,162
Thiery, C.L.	Boston	Nov. 9, 1869	96,632
		Jan. 18, 1870	99,896
		Nov. 1, 1870	108,847
		Dec. 5, 1871	121,474
		Aug. 20, 1872	130,603
		Jul. 28, 1874	153,633
Thiery, C.W.	Boston	Dec. 22, 1874	167,944
		Nov. 25, 1884	308,445
Twing, A.	Waltham	Aug. 16, 1870	106,432
Varney, N.R.	Waltham	Jan. 2, 1883	269,974
		Aug. 11, 1885	323,985
Wales, W.A.	Newton	Jan. 24, 1879	216,917
Warren, A.	Waltham	Sept 17, 1867	69,053

Wetherbee, E.D.	Waltham	Sept 1, 1885	325,296
Woerd, C.V.	Waltham	May 21, 1807	65,034
		Aug. 13, 1867	67,692
		Oct. 5, 1869	95,547
		Mar. 29, 1870	101,398
		Dec. 27, 1870	110,614
		Aug. 22, 1871	118,415
		Apr. 6, 1875	161,725
		Jul. 27, 1880	230,596
		Feb. 6, 1883	271,965
		Jun. 30, 1885	320,992
		Apr. 27, 1886	340,850
		May 11, 1886	341,786
		Dec. 7, 1886	354,002

Footnotes

1. Omitted inventions involving chains, pendants, tools and keys
2. Omitted assignments of inventions

Source: George H. Eckhardt: United States Clock and Watch Patents
(New York, 1960).

APPENDIX C
THE WATCH PAPERS OF MASSACHUSETTS

Watch papers are centuries old. They flourished because jewelers realized that they had great sales potential. These papers were used to advertise the names of watchmakers, their addresses and even other lines of business (ie they often showed jewelry items, spoons or china). The papers also served another function: they showed the dates of repairs, the name of the workman and sometimes the cost of the repair.

The original watch papers were made of fabric and were used as a cushion between the inner and the outer cases to protect the movement from dust. But, while they were first functional objects, they soon had more decorative value. Women used them to "exhibit their needlework" in their watches. (1) Watches papers were often given out at places where carriages stopped, and they were used as we now use most cards. (2)

American watch papers were not used until about 1790. Many of these were ornamental; they depicted doves. flowers. blossoms, cherubs. For this reason, they served as valentines. But in time, the functional uses of watch papers came to dominate. The designs changed to watches, clocks, watch springs, hour glasses. Father Time, and other objects that denoted or suggested time. Certainly, the most famous American watch paper was engraved by Paul Revere for Aaron Willard, around 1781. It shows Father Time, a rooster, an angel, a tree and a watch. (3)

In the list of watch papers that follows. several facts are evident. First, most watch papers preceded the growth of the American watch industry. The only notable paper in modern times is the one produced for the Waltham Watch Company. Second. the later the watch paper, the more plain--many had only the jewelers name and a slightly ornate border.

THE WATCH PAPERS OF MASSACHUSETTS

Name and Description	Date of Paper

Adams and Eaton, Boston
    Large eagle with border,
    Cream 1 color
	1817

American Watch Co., Waltham
    View of factory buildings,
    cream color
	c 1859[*]

Bacon and Smith, 4 Elm St., Boston Green paper	c 1845-53

Balch and Smith, Salem
    Large eagle holding watch,
    Cream paper
	1825

Balch, B. and Son, Salem
    Large eagle holding watch,
    Cream paper
	1834, 1843

Barrett, Humphrey, Bolton
    Scroll with leaves and berries,
    Green paper
	c 1869

Bemis, Samuel, Cambridge Watch Spring, pink paper	c 1812
Bigelow, Joseph J., 4 Elm St., Boston Cream paper	c 1830-35
Bond, Charles, 37 Washington St., Boston Cherubs holding watch	1831

Bond and Son, Congress St., Boston
    Calendar showing months in which
    the sun is faster or slower
	1840

Boyden and Fenno, Worcester
    Two cherubs holding a large watch,
    small watch above, Orange paper
	c 1842-82

Brackett, Jeffrey R., 69 Washington St., Boston   c 1838-48
    Small watch, girl with anchor, Father
    Time, white cardboard

Brooks, L.S., Essex and Amesbury St., Lawrence   c 1848-50
    8 lines of verse, pink paper

Collier, Asa C., Grafton (Medwayville)   c 1849
    Ornamental border, cream paper

Corbett, Otis, Worcester   1808
    Manuscript, cream paper

Crooks and Phelps, Northampton   --
    Name in script

Cummings, D., Barre   1853
    Ornamental Border, buff paper

Currier and Foster, Essex St., Salem   c1831
    Orange paper

Currier and Trott, 139 Washington St., Boston   1835
    Yellow paper

Cushing, P.H., Weymouth Landing   1821
    Ornamental border, cream paper

Cutler, Amos, 217 Washington St., Boston   c1834-38
    Ornamental, rose paper

Drown, C.L. and J.B., Newburyport   1841
    Ornamental, green paper

Drown, John B., Newburyport   1855, 1860
    Ornamental, yellow paper

Drown, Richard W., Newburyport   1820, 1863
    Ornamental, black on white paper

Eaton, Leander, Barre   c1844
    Ornamental, buff paper

Foster, John C., Haverhill   1827
    Large and small cog wheels, yellow paper

Fuller, A.M.B., W. Medway   c1869
    Cream paper

Gere, Isaac, Northampton Ribbons and flowers,   1801
    Red on white; black on white

Gibbs, Benjamin W. Main st. Cambridge port   1855
    Indicator showing fast and slow,
    Green paper

Glover, William, Milton   c1849
        Ornamental, white paper

Goddard, Daniel, Worcester Ornamental,   1834, 1841
        Orange paper; Yellow paper (1848)

Goddard, George S., Faneuil Hall, Boston   c1829
        Ornamental, pink paper

Gooding, John, Plymouth   1836
        Ornamental, Cream paper

Goodnow, S.H., Fitchbury   1838
        Printed, Ornamental, Yellow paper

Haynes, Joshug, Concord Woman holding scales   1835
        and watch, Cream paper

Hill, Benjamin, Cambridgeport Woman with   1815
        Watch on pyramid, Cream paper

Hobbs, Nathan, 9 Faneuil Hall, Boston   c1835
        Ornamental, yellow paper.

Howe, Jubal, Washington St., Boston   1831
        Cream paper

Huntington and Church, Springfield   --
        Name in script

James and Williams, 55 Cornhill St., Boston   1815
        Ornamental,

Johnson, J., 13 Merrimack St., Lowell   c1842
        Blue and green papers

Kettell, J.V., 32 Court St., Boston   1851
        Ornamental, cream paper

Kingman, Matthew, Woburn   1847
        Large watch in center and verses,
        Orange paper

Lamb, T.M., Harrington Corner, Worcester   c1852
        Ornamental, White paper

Lane, D.H., 52 Front St., Gloucester   c1870
        Border of type ornaments, green

Leland, Caleb, Templeton   c1810
        White paper

Macomber, A. Judson, 2 Bacon Block, Newton   c1868
        Small watch, orange paper

Morse, Moses L., Cambridgeport   1809
        Woman with watch on pyramid

Moss, (Unknown), Rochdale   1818
        Woman with Father Time

Munn and Jones, Greenfield   --
        Spray of leaves, white paper

Pope, Josiah, 36 Marlboro St., Boston   1802
        Border of blossoms and leaves,
        Cream paper

Putnam, F.H., Worcester Woman with large   c1844
        Watch, eagle holding small watch,
        Cream paper

Sawyer, George E., Feltonville   c1860
        Ornamental, brown paper (soiled?)

Sawyer, Joel, Bolton   c1880
        Girl and tree, ornamental flowers,
        Yellow paper

Sargeant, Thomas, Springfield   --
        Winged victory

Skinner, Alvah, 63 Congress St., Boston   c1830
        Ornamental, cream paper

Smith, Jesse Jr., Essex St., Salem   1841, 1861
        Ornamental leaves and flowers, buff paper

Stowell, A., Worcester   c1800
        Large eagle holding shield, two small
        watches, cream paper

Sutton, Enoch, 37, Washington St., Boston   c1830
        Eagle, orange paper

Tappan, Israel F., Manchester   1832
        Ornamental, orange paper

Terry, Geer, Worcester   1816
        Ornamental

Trott, Andrew C., 28 Marlboro St., Boston   1808
        Large eagle, cream paper

Whitney, Hiram, Watertown   c1849
        Printed, type ornaments, cream paper

Whitney, Moses, 1 Dock Square, Boston   c1830
        Cream paper

Willard, Aaron
     Angel blowing trumpet, Father Time   c1781
     Roosters, flowers, leaves, white paper

Willis, Stillman, 54 1/2 Cornhill, Boston   c1821
     Die proof impression, white paper

Wingate, James, Haverhill   c1850
     Small eagle, yellow paper

Wood, Moses, West boro   c1837
     Yellow paper

Woodworth, Earl, Springfield   1839
     Girl holding scales and watch,
     Cream paper

Woodworth and Kirkham, Springfield   c1847
     Girl holding scales and watch,
     Cream paper

Note: *C means approximate date

Sources: Dorothea Spear, "American Watch Papers," <u>Proceedings</u>, American Antigrarian Society, Vol. 61, April 18, 1951; John Ittman, "Early American Watch Papers in the Metropolitan Museum," NAWCC Bulletin, XII (Oct. 1966), 512-515.

References

1.   Dorothea E. Spear. "American Watch Papers," Proceedings, American Antiquarian Society, April 18, 1951, p. 298.

2.   Earnest A. Cramer, "Watch Papers," NAWCC Bulletin, IV (June 1951), p. 380. It should be noted that Spear informs us that E. A. Cramer has the largest private collection of watch papers in the United States.

3.   Spear, p. 305

BIBLIOGRAPHY

Books

Abbott, Henry G. *History of American Waltham Watch Company.* Chicago, 1905.

-----------. *The Watch Factories of America: Past and Present.* Chicago: G. K. Hazlitt, 1888.

A History of Technology, IV, The Industrial Revolution. Oxford: At the Clarendon Press, 1958.

American Waltham Watch Company. *A Few Points on the Construction of a Pocket Watch.* 1890.

-----------. *The Perfected American Watch.* Waltham, 1904.

American Watch Co. *Souvenir Catalogue, New Orleans Exposition, 1884-1885.* Bristol: Ken Roberts, Jan. 1972.

Averitt, Robert. *The Dual Economy.* New York: Norton, 1968.

Babson, Roger W. *Business Barometers.* Babson Park, 1937.

Bailey, Chris. *Two Hundred Years of American Clocks and Watches.* Englewood Cliffs: Prentice-Hall, 1975.

Baillie, G. H. *Watches.* Methuen, 1929.

-----------. *Clocks and Watches: An Historical Bibliography.* London: N. A. G. Press, 1951.

-----------. *Watchmakers and Clockmakers of the World.* 3rd ed.; London: NAG Press, 1951.

-----------. *Muskets to Mass Production.* American Precision Museum, 1976.

Berle, August A. and Gardner C. Means. *The Modern Corporation and Private Property.* New York: Macmillan, 1933.

Bolino, August C. *The Development of the American Economy.*

Columbus: Charles E. Merrill, 1966.

Bolles, A. S. *Industrial History of the United States*. 1880.

Boston Directory. Boston: Sampson Murdock, various years.

Brearley, Harry Chase. *Time Telling Through the Ages.* New York: 1919.

Britten, Frederick J. *Britten's Old Clocks and Watches.* New York: Dutton, 1973.

-----------. *Watch and Clockmakers Handbook.* London: SPON, 1955.

Bruton, Eric. *Clocks and Watches.* New York, 1967.

Buyers Guide to Watch, Clock Jewelry. International Publishers Service, 1979.

Calvin, Jeannette. *International Trade in Clocks and Watches.* Bureau of Foreign and Domestic Commerce, 1928.

Camm, Frederick J. *Watches.* Chemical Publ. Co., 1941.

Clark, Victor S. *History of Manufactures in the United States.* 3 vols.; Washington: Carnegie Institution, 1928.

Clutton, Cecil. *Watches.* New York: Viking, 1965. Chicago,1905.

Copeland, Alfred M. *Our Country and its People* Boston: Century Publishing Co., 1902.

Criss, David. *Collector's Price Guide to American Pocket Watches.* Imlay City, 1980.

Crossman, Charles S. *The Complete History of Watchmaking in America.* Adams Brown, Reprint of Jewellers Circular and Horological Review, 1885-87.

Cumhaill, P. *Investing in Clocks and Watches.* New York: Chicago, 1905.

Cuss, Theodore C. *Cammerer Cuss Book of Antique Watches.* Suffolk:

Baron Publ., 1976.

-----------. *The Story of Watches.* 1952.

Cutmore, M. *Watch Collector's Handbook.* Rutland: Charles E. Tuttle, 1976.

Daniels, George. *English and American Watches.* London: Abelard Schuman,1967.

Dunwell, S. *The Run of the Mill.* David Godine. 1978.

de   Carle,   Donald.   *Watches   and   Their   Value.*   London: N. A. G. Press, 1978.

Dixon, J. *Watch Companies of America.* 1978.

Dyer, George Lewis.   *The Story of Edward Howard and the First American Watch.* Boston: E. Howard Watch Works, 1910.

E and J Swigart Co., *Catalogue.* 1911.

Eckhardt,   George   H.   *United   States   Clock   and   Watch   Patents, 1790-1890.* New York, 1960.

Edwards, Bernard J. *Watch and Clock Advertising.*   Chicago: Review Advertising, 1976.

Edwards, John (Comp.).   *The Complete Checklist of American Clock and Watchmakers, 1640-1950.* Stratford: The New England Publ., 1977.

Ehrhardt, Roy.   *American Pocket Watch Price Indicator.*   Kansas City, 1980.

-----------.   *American Pocket Watches, Encyclopedia and Price Guide.* Kansas City: Heart of America, 1982, Vol. I.

-----------.   *Foreign and American Pocket Watch Identification and Price Guide.* Kansas City: Heart of America Press,1976

-----------.   *Pocket Watch Price Guide.*   Kansas City: Blue Ridge Extension, 1972.

-----------. *Waltham Pocket Watch Identification and Price Indicator.* Kansas City, 1976.

Equity Watch Co. *The Equity Watch.* Boston, nd.

Etchells, C. T. *Repairing Repeating Watches.* 1917.

Fredyma, M. C. and P. J. *A Directory of Boston Silversmiths and Clock Makers.* 1975.

Fried, Henry B. *Cavalcade of Time.* Dallas: Zale Corp., 1968.

-----------. *The Watch Escapement.* New York: Jadow Publ., 1959

Gibbs, James W. *The Dueber-Hampden Story.* Phila., 1954.

Goss, Elbridge H. *Bibliography of Melrose.* Williams, 1889.

-----------. *The History of Melrose.* Melrose, 1902.

Green Brothers. *Catalogue.* New York: Maiden Lane.

Greenough's Directory. Melrose, 1974-79.

Hampden Watch Company. *Illustrated Catalogue and Price List.* 1892.

Hays, H. Martin. *Collectors's Guide to American Pocket Watches.* 1976.

Harris , H. G. *Collecting and Identifying Old Watches.* New York: Emerson, 1978.

Harrold, Michael C. *American Watchmaking: A Technical History of the American Watch Industry, 1850-1930.* Supplement, NAWCC, Bulletin, Spring 1984.

Hawkins, Vernon M. (Transcriber). *Annual Reports to Stockholders, American Waltham Watch Company, 1859-1899.* West Boxford, 1984.

Hilker, Emerson W. (Compiler). *Horological Books and Pamphlets in the Franklin Institute Library.* 3rd ed. Phila.: The Franklin Institute, 1974.

Hilton, George W. and John F. Due. *The Electric Interurban Railways.* Palo Alto: Stanford University Press, 1960.

Historical Records Survey. *Index to Local News in the Hampshire Gazette, 1786-1937.* 1939.

-----------. *Inventory of City and Town Archives of Massachusetts.* 1939.

-----------. *Guide to Manuscript Collections in the Worcester Historical Society.* 1941.

-----------. *Publications of the Waltham Historical Society.* 1919-36.

Howard, E. and Co. *Clocks and Fine Watches.* N. E. Publ., 1976. Reprint of 1858 edition.

-----------. *Catalogue.* 1909.

Howard Watch & Clock Co. *Illustrated Catalogue.* Boston: Tchelder & Wood, 1871.

Kendal, James Francis. *A History of Watches.* London: Crosby Lockwood and Son, 1892.

Landes, David S. *Revolution in Time.* Cambridge: Belknap Press, 1983.

Le Coultre, Francois. *A Guide to Complicated Watches.* Bienne: Charles Rohr, 1952.

Lockwood, John Hoyt. *Western Massachusetts: A History, 1636-1925.* New York: Lewis Publishing Co., 1926.

Lothrup's Melrose Directories

Martin, Thomas P. *History of the Dennison Manufacturing Company.* Framingham, 1922.

Massachusetts Register and Directory. Boston: Samson Davenport, 1858.

Mayr, Otto and Robert C. Post (eds.). *Yankee Enterprise: The Rise of the American System of Manufacturers.* Washington: Smithsonian Institution, 1982.

Milham, Willis I. *Time and Timekeepers.* New York: Macmillan, 1923.

Minnesota Watchmakers Association. *American Watch Historical Information with Serial Numbers and Dates.* Champaign: Parkland College Library, nd.

Moore, Charles W. *Timing a Century--History of Waltham Watch Company.* Cambridge: Harvard University Press, 1945.

Morrow, James B. *The Man Who Holds a Watch on 125,000 Miles of Railroads: Webb C. Ball.* 1910.

Muhr's (H) Sons. *Columbian Exhibition.* Philadelphia, 1893.

NAWCC. *Pocket Time Pieces of the New York Chapter Members.* New York, 1968.

New York Watch Co. *Serial List.* 1872-1875.

North, S. N. D. and Ralph North. *Simeon North, First Official Pistol Maker of the United States.* 1913.

Niebling, Warren. *History of the American Watch Case.* Philadelphia: Whitmore Publ., 1971.

Ohlson, Olaf. *Helpful Introduction for Watchmakers.* Waltham Watch Co., 1928.

Palmer, Brooks. *The Romance of Time.* New Haven: Clock Manufacturers Association of America, 1954.

Rosenberg, Charles. *American Pocket Watches.* 1965.

Sanderson, Edmund L. *Waltham Industries.* Waltham: Waltham Historical Society, 1957.

Shugart, Cooksey. *The Complete Guide to American Pocket Watches.* Cleveland: Overstreet, 1981.

Stelle, J. Parish. *The American Watchmaker and Jeweller.* New York: Jesse Haney, 1868.

Stone, Orra L. *History of Massachusetts Industries.* Boston, 1930.

Thomson, Richard. *Antique American Clocks and Watches.* Princeton: Van Nostrand, 1968.

Townsend, George E. *Almost Everything You Wanted to Know About American Watches and Didn't Know Who to Ask.* Vienna, Va., 1970.

-----------. *American Railroad Watches.* Privately Published, 1977.

-----------. *E. Howard & Co.* Kansas City: Heart of America Press, 1982.

Trouble Shooting, Adjustment, and Repair. Arco Publ. Co., 1948.

Ullyett, Kenneth. *Watch Collecting.* Chicago: Regnery, 1970.

U. S. Centennial Commission. *Reports and Awards.* VII.

U. S. Department of Commerce. *Census of Manufactures: 1927, Clocks, Watches and Parts.* 1929.

U. S. National Recovery Administration. *Code of Fair Competion for the Assembled Watch Industry.* Washington, 1934.

Vatter, Harold G. *The U. S. Economy in the 1950's.* New York: Norton, 1963.

Waltham Precision Instrument Co. *American Waltham Watch Co., 1885 Trade Catalogue.* Bristol: K. Roberts, 1972.

Watchmaking in America. Robbins, Appleton and Co, 1870.

Wilkinson, T. J. *Escapement and Train of American Watches.* Keystone, 1928.

Ziebell, R. J. (Compiler) *Serial List: New York Watch Company,* Springfield: Old Post Office Clock Shop, 1972.

Articles

Aked, Charles K. "The First English Watch Manufactory, NAWCC

Bulletin, XXVIII (Feb. 1986), 52-60.

American Horologist

"American Watch Co., Waltham Mass.," New Orleans Exposition, 1884-1885 (1972).

Amend, Adolph Jr.   "Interesting Construction Features of Early Howard Watches," NAWCC *Bulletin*, XIX (April 1977), 154-55.

Autry, Peyton, "Waltham's Magical Maximus," NAWCC *Bulletin*, XXII (Oct. 1980), 531-539.

Babcock, Harrison F. and Mary E. "The Railroad Watch," NAWCC *Bulletin*, IX (Oct. 1960), 385-387.

"Ball by Waltham," NAWCC *Bulletin*, (April 1976), 224.

Battison, Edwin A. "The Auburndale Watch Company". Museum of History and Technology, *Bulletin 218*, (1959).

"Besieged Industry Plans New Strategy," *Business Week*, Sept. 20, 1952, 38-41.

Blackwell, Dana J. "Answer Box," NAWCC *Bulletin*, XXVII (Oct. 1985),605-606.

-----------.   "Answer Box," NAWCC *Bulletin*, XV (Feb. 1973), p. 999.

Bolino, August C. "The Elusive Model A," NAWCC *Bulletin*, XXIV (April 1982), 167-168.

-----------. "Who (Or What) Killed Waltham, NAWCC *Bulletin*, XXV (Oct.1983), 555-557.

Borg, Arthur N. "The Howard Ten Size Watch," NAWCC *Bulletin*, XII (Aug. 1967), 941-964.

Boston Watch Company. Complete Papers, Jan. 13, 1934 to April 1, 1939, *Bulletins*, Special Papers.

Brown, Joseph E. "The Invention of an Improved Compensation

Balance: The Genius of Charles Vander Woerd," NAWCC *Bulletin*, XVIV (Oct. 1982), 487-493.

Burt, Edwin B. "Waltham Dial Company," NAWCC *Bulletin*, VII (April 1956), 123-128.

Carosso, Vincent P. "The Waltham Watch Company," *Bulletin of the Business Historical Society*, XXIII (Dec. 1949), 165-189.

Coleman, J. E. "Clocks, Watches Collectors, People, and Legends," NAWCC *Bulletin*, XI (Oct. 1964), 426-432.

Conley, Robert N. "Waltham: A City That Became a Pocket Watch." NAWCC *Bulletin*, XXIII (Oct. 1981), 481-483.

Cramer, Ernest A. "The Watch Case and the Watch, NAWCC *Bulletin* VI (Oct. 1955), 552-562; VII (Feb. 1957), 397-411.

-----------. "Watch Papers," *NAWCC Bulletin*, IV (June 1951). 380-81.

Dean, Joseph. The Trement Watch Company," NAWCC *Bulletin*. IX (Feb. 1960), 158.

De Fazio, Thomas L. "Ball & Company and the Vanderbilt Railroads," NAWCC *Bulletin*, XVIII (October 1976). 411-419.

-----------. "The American Waltham Watch Company Repeaters." NAWCC *Bulletin*, XIX (June 1977), 271-278.

-----------. The Nashua Adventure and the American Watch Company." NAWCC *Bulletin*. XVII (December 1975). 575-589.

Doan, Edward N. "I Didn't Know That," NAWCC *Bulletin*, XIV (Oct. 1971), 1397-1404.

Dodge, Albert O. and George G. Lucchina, "Jonas G. Hall, 1822-1891," NAWCC *Bulletin*. XVIII (Oct. 1976). 436-452.

"Eighty Years' Progress of the United States: A Family Record of American Industry, Energy and Enterprise." 1868, Reprint, NAWCC *Bulletin*, August 1981. 369-373.

"Family Quarrel," *Business Week*, April 20,1946, 31-32.

Forgie, Fraser R. "The Boston Watch Company," NAWCC *Bulletin*, XIII (Feb. 1968), 143-45.

Fuller, Eugene T. "U. S. Watch Company Centennial Commemoratives," NAWCC *Bulletin*, XIX (Oct. 1977), 494-501.

Gibbs, James W. "Sucker Bait---(Horological, not Piscatorial)," NAWCC *Bulletin*, IX (Aug. 1961), 791-797.

----------. "Watch Hands: The History of J. H. Winn Inc.," NAWCC *Bulletin*, XI (Apr. 1964.), 224-227.

----------. "Who Was America's First Watchmaker?" NAWCC *Bulletin*, XVIII (Dec. 1976), 556-565.

Gohl, Anthony. "The Wrist Watch." NAWCC *Bulletin*, XIX (Dec. 1977).

Gonzales, F. A. R., "Answer Box," NAWCC *Bulletin*, XII (Feb. 1961), 158.

Gould, R. E. "The Testing of Timepieces," U. S. Department of Commerce, *Circular C*, 432, 1941.

Hafter, Daryl M. "The Business of Invention in the Paris Industrial Exposition of 1806." *Business History Review*, XVIII (Autumn 1984).

Hauptman, Wesley R. "America's Half-Breed Watch," NAWCC *Bulletin*, XII (Apr. 1966), 270-275.

-----------. "An Incredible Reunion," NAWCC *Bulletin*, Xi (Dec. 1964), 555-563.

-----------. "Appleton, Tracy and Company," NAWCC *Bulletin*, X (Apr. 1963), 690-700.

-----------. "Serial Numbers of the First Waltham Watches," NAWCC *Bulletin*, VIII (Oct. 1958), 259-62.

-----------. Swiss Imitations of Early American Watches," NAWCC

Bulletin, IX (Aug. 1960), 270-276.

-----------.   "The American Watch Company," NAWCC *Bulletin*, XI (April 1964), 170-209.

-----------.   "The Boston Watch Company," NAWCC *Bulletin*, X (Oct. 1963), 923-940.

-----------.   The First Waltham. . .Or Are They Howards?"   NAWCC *Bulletin*, X (Dec. 1961), 27-36.

Hoke, Donald.   "Ingenious Yankees, The Rise of the American System of Manufactures in the Private Sector, *Journal of Economic History*, XLVI (June 1986}, 489-491.

Hounshell, David A. "The System: Theory and Practice," in Otto Mayr and Robert C. Post (eds.).   *Yankee Enterprise*. (Washington: Smithsonian Institution Press, 1981.

Hovey, H. C. "The American Watch Works," *Scientific American*, August 16, 1884, 101-104.

Howard, E. "American Watches and Clocks," in Chauncey M. Depew (ed.). *One Hundred Years of American Commerce*.   New York: D. O. Haynes, 1895.

"Howard Watch Identification," NAWCC *Bulletin*, XX (June 1978), 284.

Hughes, Lawrence M. " Who Killed Waltham?"   *Sales Management*, April 15, 1950, 37-39; 116-126.

"International Trade in Clocks and Watches."   *Trade Information Bulletin*, No. 584, U. S. Department of Commerce,

Ittman, John.   "Early American Watch Papers in the Metropolitan Museum," NAWCC *Bulletin*, XII (Oct. 1966), 512-515.

Jackson, B. J. "Imitation Key Wind Watches," NAWCC *Bulletin*, XXI (August 1979), 426-434.

Jefferson, J. George.   "Another Howard (Waltham Product) Watch," NAWCC *Bulletin*, XIII (Oct. 1968), 548.

Jewelers Circular, 1867-1977.

Jewelers Weekly. 1887-1899.

Kalish, Charles. "My Favorite Watches." NAWCC *Bulletin.* IX (Dec. 1960), 473-495.

Kleeb, Alvin A. "Watch Jewels of the Past." NAWCC *Bulletin.* X (Apr. 1962), 191-198.

Landes, David S. "Watchmaking: A Case Study in Enterprise and Change," *Business History Review,* LIII (Spring 1979), 1-39.

Lavoie, Robert A. "Keystone Howard Series 3 or 9?" NAWCC *Bulletin* XXVIII (Feb. 1986), 81.

Lindberg, Steve. "The Production History of the Waltham Maximus," NAWCC *Bulletin,* XXVII (Apr. 1985), 174-188.

Livingstone, Gordon A. "The Hampden Watch Company," NAWCC *Bulletin,* XIV (Feb. 1970), 155-158.

"Many Watches," NAWCC *Bulletin,* XI (Oct. 1965), 986-989.

Manufacturing Jeweller. 1884-1959.

Massey, Mark, "When May a Watch be Considered an Antique," NAWCC *Bulletin,* III (May 1948), 417.

Mathews, A. E. "A Crystal-Plate Waltham Watch," NAWCC *Bulletin,* XIV (April 1971), 1075-1077.

-----------. "American Watch Sizes," NAWCC *Bulletin,* XIV (Aug. 1970), 434-437.

-----------. "Answer Box," NAWCC *Bulletin,* XXII (Aug. 1980), 412.

Meggers, W. F. "The Railroad-Marked Watch." NAWCC *Bulletin,* XXIV (February 1982), 3-49.

Miether, William E. "The Babcocks--Mary and Harrison--and Webb C. Ball," NAWCC *Bulletin,* XI (Oct. 1964), 439-446.

Moore, C. W. "Some Thoughts on the Early Labor Policy of the Waltham Watch Company," *Bulletin of the Business Historical Society*, XIII (April 1939), 25-29.

Mulvin, Walter J., "Two Sleepers," NAWCC *Bulletin*, VII (June 1956), 226-227

National Jeweller. 1906-1968.

Niebling, Warren H. "A History of American Watchcase," NAWCC *Bulletin*, 4 parts, XIV (Dec. 1969), 28-29; XIV (Feb. 1970) 167-171.

-----------. "The Hobby, the Movement and the Case," NAWCC *Bulletin*, XIII (Dec. 1968), 627-634.

Nijssen, Gerrit A. "The American Repeater," NAWCC *Bulletin*, XVI (December 1973), 4-22.

Olson, David C. "The Auburndale Rotary," NAWCC *Bulletin*, VIII (April 1959), 509-510.

Pease, Hamilton E.. "Imitations of American Watches." NAWCC *Bulletin*, IX (Feb. 1961), 709-710.

Peghiny, James W. "Some Interesting Photographs of the Auburndale Watch Company," NAWCC *Bulletin*, XIV (Dec. 1970), 790-794.

Porter, Robert D. "Waltham's Up and Down Work," NAWCC *Bulletin*, XXVII (October 1985), 577-581.

Pritchard, Mrs. W. L. "Words from the Past," NAWCC *Bulletin*. XVIII (Dec. 1976), 598-603.

Roe, Joseph W. "The Colt Armory," in *English and American Tool Builders*. New Haven, 1916.

Rosenberg, Charles, "Pitkin: Fact or Fiction," NAWCC *Bulletin*. X (Feb. 1963), 582-590.

"Running Down Watch Company Gets a New Mainspring," *Business Week*, June 22, 1957, 111.

Selchow, Frederick Mudge. "The Watch Company of Fitchburg Massachusetts," NAWCC *Bulletin*. XIII (Apr. 1969), 914-922.

-----------. "1812 Belding Dart Bingham 1878; 1859 The Nashua Watch Co., 1862," NAWCC *Bulletin*, XVII (Dec. 1975), 539-574.

Settle, Steven D. "Home Watch Co.," NAWCC *Bulletin*, XIX (Apr. 1977), 207.

Small, Percy Livingston. "Don J. Mozart," NAWCC *Bulletin*, VII (Apr. 1957), 466-471.

-----------. "Luther Goddard and his Watches," NAWCC *Bulletin*, V April 1953) 355-363.

-----------. "Napoleon Bonaparte Sherwood," NAWCC *Bulletin*, VI (Dec. 1953), 24-28.

-----------. "The Pitkin Brothers," NAWCC *Bulletin*, VI (Oct. 1954), 251-260.

Small, P. L. and F. E. Hackett. "E.Howard: The Man and the Company," NAWCC *Bulletin*, Supplement, 1962.

-----------. "The Lancashire Gage," NAWCC *Bulletin*, V (April 1952(, 148-149.

Spear, Dorothea E. "American Watch Papers." *Proceedings*, American Antiquarian Society, April 18, 1951.

Stephens, Carlene E. Speech Delivered Before a Meeting of Chapter 12 of the National Association of Watch Collectors, Lanham, Maryland, October 10, 1984.

Stevens, W. B. " A Group of Trusts and Combinations," *Quarterly Journal of Economics*, XXVII (August 1912), 602-608.

Stephens, W. Barclay. "The Adventure of the Lord's Prayer," NAWCC *Bulletin* VI (Apr. 1954), 130-138.

-----------. "The New York Watch Company," NAWCC *Bulletin*, IV (Feb. 1951), 279-287.

Summar, Donald J. "Columbia, Suffolk, United States and Betsy Ross
     0-Size Movements," NAWCC *Bulletin*, XXVI (Dec. 1984),
     717-719.

"Swiss Watches and U. S. Trade," *Business Week*, Dec. 26, 1953, 70.

The Keystone. 1879-1934. (Merged with *Jewellers Circular*)

"The Priceless Possession of the Few," NAWCC *Bulletin*, Supplement,
     Winter 1974.

"Time's Turning," *Business Week*, Sept. 8, 1945, 54-55.

Townsend, George E. "Answer Box," NAWCC *Bulletin*, XXIII (Aug.
     1981), 410.

Treiman, Larry. "Railroad Watches and Time Service," NAWCC
     Bulletin, 651-675. See also the "Answer Box," NAWCC
     Bulletin, XVIII (April 1976), p. 224.

-----------. "Timing the Trolleys," NAWCC *Bulletin*, XX (Feb. 1978),
     3-19.

Tremayne, Arthur, "The Introduction of the Wrist Watch," NAWCC
     Bulletin, II (Aug. 1947), 265.

U. S. Department of Commerce. *Trade Information Bulletin*, No. 584
     (1927).

U. S. War Department. "Wrist Watches, Pocket Watches, Stop
     Watches, and Clocks," *Technical Manual 9-1575*, April 6, 1945.

"Waltham and the Importers," *Newsweek*, (Feb. 14, 1949), 64.

"Waltham Opens Again," *Newsweek*, Aug. 28, 1950, 63.

"Waltham Windup," *Business Week*, Mar. 15, 1952, 36.

"War's Timepieces, *Business Week*, Apr. 4, 1942, 67-68.

White, George V., "The Early American Watch--Waltham," NAWCC
     Bulletin II (Aug. 1947), 279-283.

Williams, Robert G. "Some Notes on Modern Railway Timekeeping,"
NAWCC Bulletin, XVI (Feb. 1973), 977-980.

Williamson, R. C. "Military Timekeepers," NAWCC *Bulletin*, XIV
(April 1970), 251-261.

Newspapers

New York Times

Oct. 2, 1868, "The Waltham Watch." p. 2.

July 20, 1879, "Statistics of Watch Manufacture." p. 10.

Aug. 16, 1882, "Manufactures of Different Countries Compared," p.
4.

Jan. 26, 1888, "Erie Railway Order Inspection," p. 4.

Sept. 2, 1888, Smallest Watch in America." p. 12.

May 26, 1903, "Watchmakers in a Deal," p. 16.

July 26, 1903, "Watch Trade Trust," p. 11.

Aug. 9, 1912, "Abolish 20 Year Guarantees," p. 5.

May 30, 1921, "Swiss Watch Industry Facing Crisis," p. 6.

Dec. 10, 1927, "Smuggling Swiss Watch Movements." p. 19.

Oct. 19, 1934, "Nationwide Smuggling Racket," p. 46.

June 22, 1944, "WPB Lists Watch Shortages." p. 25.

Aug. 9, 1947, "Watch Prospects Good," p. 24.

Feb. 3, 1949, "77 % of Watches Sold are Swiss," p. 12.

Boston Post

Fitchburg Daily Sentinel

Melrose Advertiser

Melrose Reporter

Waltham Sentinel